T0196306

Omar's Guide for Surviving this Turbulent Age including Poetry and Thoughts to Consider

Omar's Guide for Surviving this Turbulent Age including Poetry and Thoughts to Consider

Topics and Poetry for Your Consideration

OMAR

Omar's Guide for Surviving this Turbulent Age including Poetry and Thoughts to Consider
TOPICS AND POETRY FOR YOUR CONSIDERATION

iUniverse books may be ordered through booksellers or by contacting:

iUniverse
1663 Liberty Drive
Bloomington, IN 47403
www.iuniverse.com
1-800-Authors (1-800-288-4677)

ISBN: 978-1-4917-8763-2 (sc)
ISBN: 978-1-4917-8764-9 (e)

Library of Congress Control Number: 2016900760

Print information available on the last page.

iUniverse rev. date: 1/12/2016

Contents

To my wife, for giving me 53 glorious years of her beautiful life, being my happiness, building our family, insulating me with her shining eternal love. She shall greet me at the gates of heaven.

1.0 Author's Prologue to the Reader

The author's views as expressed in this handbook are meant to be practical, honest, common sense solutions to all levels of problems: personal, family, professional, governmental, and international. In all such cases only common sense approaches, with solutions are offered, on the topic being read.

The poetry section contains deep reflections upon aspects of life every human being encounters. Perhaps, if prose is not enough to guide you in survival, the poetry will help. The author certainly hopes that its added deep dimensions to the human inquiry of life shall illuminate solutions to the reader's needs in our "constructive destruction" way of life we find ourselves immersed in this turbulent century.

Hopefully the reader may find some humor in the latter section called "Helpful Guide to Omar's Manifesto", as he claims to be a resurrected man some 3000 years ago from the biblical lands, who communicates his proposed solutions to survival as if he were in our present year. The section contains much conjectured information on Omar himself, background, sayings, etc ... Enjoy, the topics for thought and poems for meditation.

1.1 General Discussion

"A Handbook: Surviving This Turbulent Age" is a book of suggested common sense solutions on over coming the constructive destruction pressures being exerted upon the following: individuals, families, organizations, entire societies, governments, etc ... Such pressures are being experienced today, and perhaps for the rest of our 21st century. Such pressures are a threat to world peace, and basic human rights, that all of humanity should and must have to thrive, advance, and even to survival itself.

Life, liberty, and the pursuit of happiness around the world is indeed at risk. In other words, how do we survive during a period of such historic change, never seen before, that is occurring at such an unbelievable rate.

Such pressures are brought about because of several factors such as: population growth, instant communications, ease of global travel, global commerce, global investment, capitol availability, internet availability, global linking of educational institutions, cultural exchanges, etc ...

Because of the above factors, many entrenched groups capable of wielding resistance to change have been, and shall continue to be disruptive to such positive change, as our world societies become more fluid and common.

Unfortunately, religious differences seem to inspire the most destructive resistance of all. Indeed, religion intolerance endangers all of our global population at this time.

It is fanned by radical holy men who seemly relish the power they achieve by quoting their beliefs to easily lead gullible weak minded people. Such people seem to reject all rational common thought, completely forgetting solid common sense, and rejecting acceptable civilized behaviors. Instead, they substitute the ravings of the radicals, who claim they represent God. Such nonsense leads to total destruction of all worthy human values.

1.2 The Nom De Plum (for humor, and relief of deep thoughts)

The Nom De Plum of the author, "Omar the Enlightened One", was the final selection after many trials by the author. Again note that Omar's description changes often in this document, depending upon his, Omar's, whim of circumstances.

After all, Omar is a noted scholar, scribe, historian, philosopher, poet, engineer, designer, adviser, warrior, etc ... (you name the field, Omar has been there and done it), because he has spent much effort in different path's that life offers (perhaps, but not probable at all) on this "good earth" trying to direct people to make better choices during turbulent times.

Omar may have lived in ancient times some 3,000 years ago during King David's reign as King of Israel. Or, Omar may have lived in several different time periods. But, Omar is here, right now, in the 21st century as a fact.

You don't believe it? Just read on, my skeptical friend!

After all, archival evidence exists, found and translated, by some of the best experts in the linguistic fields during "digs" (most keep secret, for obvious reasons) in various parts of the world, and verified by carbon dating down to the year of its creation. Some biblical scholars (not to be identified) have translated Omar's earliest writings found at Israel "dig's" sites, and have been carbon dated them back to King David's time.

Indeed, other supposed Omar documents have been tied to different time periods of history as we know it.

Poetry, similar to Omar's (which is similar to King David's in Psalms), is abundant among our known greatest poets, some experts say. It is thought by some that Omar was a scribe in King David's service.

Astounding, is it not?

Other experts say his poems, and written thoughts, on various subjects remind them of Joyce, Bacon, Puskin, Dickinson, etc … Omar's style is very close to King David's, the author of Psalms in the Old Testament, so perhaps the evidence shows that he was an adviser to King David, and perhaps to King Solomon as well.

Who really knows?

Anyhow, see the table of contents of this book, and see what Omar wrote on your particular topic of interest, and see if his thoughts on such subjects are helpful to you. I would think his solutions and yours in comparison are very similar, workable, and based on good common sense.

After all, humans beings have changed very little since recognized civilizations have been established well over 5,000 years ago.

Indeed, nothing is "new under the sun", as far as human behavior is concerned.

1.3 Top Down

This guide for our turbulent age may be read from its beginning (top down) for continuity of its presentation of survival approaches to each topic's perceived problem.

This is the suggested approach, as it will expound upon, elaborate upon, and clearly offer solutions to each perceived problem under discussion. The reader may or may not like the proposed solution. Good! The author wants the reader to think.

1.4 Bottom Up

This guide may be read from its ending forward (bottom up) for its explanation of terms and definitions used in the author's musings on survival problems we all face either as individuals, or in our organizations, and our society as a whole body. It is suggested that the sections on Terms and Definitions be referenced first to aid in the readers understanding.

Then the top down read is suggested and probably should be done first.

However, poetry lovers may want to read section (5.0) first, for its beautiful messages, which touch on many deeper issues felt by all during our living in turbulent times fraught with constructive destruction examples occurring in our economy every day.

1.5   Historical View

Historical background is given, as required, to clarify the problems that must now be dealt with, by using the author's suggested survival techniques. Almost all are simple tried and true common sense found by all reasoning civilized people in their own experience's in life.

However, our current environment opposes a true common sense approach with much stupidity, stubbornness, and selfish nonsense. Omar calls this the "S-Cubed" mind set.

Just look at the previous 40 years or so actions of the US Congress. A perfect illustration of the "S-Cubed" mind set.

1.6   Omar's Simple View

Simple common sense based, upon mankind's better experiences, found through at least five to ten thousand years of attempts at civilization of society as a whole, by applying lessons learned, composes Omar's "Simple View" in this survival guide. However, the key word here is "applied".

Vested interests, politics, power seekers, easily mislead minds, non-thinking individuals, alteration of facts, etc ... and many other factors than can be listed in this survival guide are being, and certainly have been, brought in order to kill common sense solutions.

2.0 Omar's Poetic Comments

Poetry is not read by enough people today for various reasons, including one that most modern poets are not very good at their craft. Also, poetry must be read in a non-hurried fashion, and its true meaning carefully sensed, and felt by the reader. Omar admits that he is now developing his approach to prose and poetry.

All poems are original and written by Omar contained in this guide. Hopefully their meaning will be understood with clarity and meaning by reader.

See the contents of sections 5 and 6 of this survival guide for Omar's poetry.

Poetry (or something similar) is probably the language used by God and His angels, as it conveys meaning of spirit and soul, in the deepest eternal fashion, between our eternal living forces. However, it must be read and sensed properly to fully savor its meanings.

Poetry is to be felt, very deeply, in the very core of its reader. Deep emotions must be conveyed in this manner, as they cannot be in prose, no matter how well written.

3.0 Omar's Comments On:

Each following topic should be read in order to get the full gusto and suggested action being conveyed in this survival guide document to its careful reader by Omar.

In short, each subject's topic is useful and essential in today's very turbulent environment.

A useful thought can cause proper action, with sought after results.

3.1 Philosophy

Much of written philosophy is plain bull. Piles of manure. Words expressing to its reader confusing nonsense.

An accepted philosophy should be a guide to living, containing rules to follow, examples resulting from application of the rules, and be full of common sense.

Unfortunately philosophy-spouting idiots abound, and they do not really think, but hid behind their many diplomas. Much of our very expensive university level "education" is wasted in this fashion on pseudo philosophy.

For example what good is "Black Studies"? Or much of "Political Science" either. Where is reality found in such? Where is any philosophical value found in such? "Black Studies" is just part of history, unless one wished to study every tribe originating on the African continent. And do so in great detail.

"Political Science" is the art of lying, conniving, or worming your way into power in your societies structure. And then keep in power as long as possible by any means. Of course this applies also today in many corporation's management structures, especially in those deemed "To-Big-To-Fail".

Examples are General Motors, Bank of America, CTI, J.P. Morgan Investments, etc … There is no philosophical value found in such modern corporations. None at all. Their task is to make money with a profit, using every means possible, but not to produce a philosophy, except a philosophy following Gordon Gecko's famous quote, "Greed is good!". Another part of that may be expressed in the quote: "Anything goes that you can get by with.".

Today much of the 100 Great Literatures in Books by authors such as Shakespeare, Bacon, Scott, Millville, Dumas, Voltaire, Confucius, etc ... if read will offer many lessons in applied philosophy, but unfortunately, are omitted in most scholars education and quest for a degree that sells the holder of same.

Therefore, educators should require some minimum in our high schools in English, government (civics), and history classes at least touch on the major thoughts and works generated by philosophers of note : Perhaps the list should include some of the following: Pythagoreas, Socrates, Plato, Aristotle, Machiavelli, Confucius, Hume, Descartes, Thorpe, etc ... depending upon the philosophical field (logic, metaphysics, ethics, political, religious, applied, etc ...) at least two from each major philosophical field be inserted into the curriculum.

Common sense for proper survival in turbulent times seems to require such action.

3.2 Life's Purpose

Probably for the past 10,000 years or so human beings have been trying to understand what life's purpose is all about,. Of course breathing (air), drinking (water), shelter (caves), clothing (animal skins), eating (food of any kind), and finally reproductions (sex of course) came first, before time for even thinking occurred.

Then the question: "What is Life's Purpose, or perhaps better asked, "Just why are we here?" or "Why do we want to survive?", etc …

The answer is very simple. God, our Creator, made it a central feature of our soul, imbedded into the center of our spiritual self, and that life's purpose is meant to be a growth experience for each human being, as part of our eternal self.

Ancient Greeks such as Socrates, Plato, Aristotle, and philosophers in many countries (China, India, Egypt, etc …) through out history have pondered mankind's purpose, with of course always the expected disagreements, since we are the most argumentative species probably in the Milky Way Galaxy.

Throw in the topic of religion (covered in a following paragraph of this survival guide) and the topic gets rather dicey, and even dangerous. So be it.

3.3 History and Education

Historians are very rarely truthful, accurate, or fully correct. History is just a snapshot of some small period of time, or some happing in a society, or a brief rendition trying to capture a slit of the timeline at a particle place.

The same set of occurrences, as remembered by historians, are presented very differently. Facts are misrepresented, conversations not reported, and outcomes flavored by the "victors" become history. So, one must be careful as we do rely on historical works to help predict survival paths to take, and compile mass statistics for guidance, compute probabilities, and weigh options to take.

Even though history is probably one of the best guides we have for survival guidance, flawed as it is, today's educational seems to either ignore it or pass it by as a topic to be taught correctly. For example, few of our college students, much less college attendees, in our school systems today even know when World War II started, or what participated such a bloody century as the just passed 20th. And why it's aftermath is still with us.

This is a shameful comment upon our society. It is bad enough that the subject of algebra causes such academic fright, but history? In order to survive a person must be educated in today's global environment, and apply that knowledge, to make himself, family, country, region, and world a better place for all of humanity.

Educational needs covering a life time for all professions, trades, and skills in this globalize world we are now part of are available for use by everyone, as long as they are kept available, and affordable, by governments, educational groups, communications, corporations, etc. ...

Unfortunately, in the United States, a proper education has become too costly for most needing same. Greed, even in our "higher institutions of learning", is now rampant.

3.4 Religion

The most explosive topic now, and has been throughout history, is that of religion. It has been used to gain power over the masses, enslave peoples, and even to condoning justifiable genocide. Religion has been sought since mankind lived in caves, before civilization itself began, and was present as our race was learning to think more than other animals.

Religion is tied together with all other of humankind's endeavors from the very beginning of our existence on this SOL3 planet we have named Earth. Today it is being used again by one of the great Abrahamic faiths (Islam) to justify by some crazed power seekers control over great swaths of Africa, Asia, and soon other areas of our world, in a crusade of terror. Vile acts of unspeakable horror are being justified in the name of religion.

The leaders of such damage are snuffing out lives, including women and children, without mercy of any kind. They are criminals of the worst sort. Even if such damage were totally stopped today, it will take the rest of this century to erase and recover from same. If it can be stopped at all.

Unfortunately, other religions have done the same over our globe during past centuries and millennial. All under Allah's, God's, Buddha's, Vishnu's, etc … blessings according to the particular religion's leader which willing weak-minded killers are following at the moment.

Religious leaders are lacking even in our United States today. There is no movement that I see for going to our roots of common sense Christian practices as preached by such men as Billy Graham. Even the Roman Catholic leadership has stumbled very badly over the issue of pedophilia in the priesthood, and has hung itself up politically speaking on the issue of rights that a women needs over her own body.

Guidance to survival in religion is one of common sense. Or it should be. Are people so week-minded that a person who says "I am holy, know God's message, so stop thinking, drop your common sense, do what I say, I am nearer to God, etc …" is closer to God our Creator than each of us? What nonsense!

We, you and me, are each equal in God's view. In the final analysis, each individual is responsible directly to God for all his actions done in the individual's life, from birth to death. No one has spiritual authority between each of us and God Himself.

Each individual has his own book of life, as he lives it, and to be held accountable for as death takes him. Each one of us will be held accountable.

Live life to the fullest while here, but be very careful of "holy" men.

3.5 Today

Today the entire SOL3 is globalized with radio, telephone, satellite communications, Internet, the Webb (supported by the Internet), computerization of everything possible (more yet to come), transportation systems(sea and air), very easy movement of people, cultural emergence, race intermarriage, etc …

We are one interdependent world, and we all must get used to same, as it will never go back to what it has been even in the recent past. Being able to survival has just been expanded in such a world.

Complexities abound in all facets of living more than ever before in a globalize world. Leaders at all levels are having to understand the new realities. Some cannot adjust to the new realities.

Religious leadership, politicians, and other such areas are much further behind than technology, science, engineering, and military leadership in the art of change as driven by globalization.

Each individual's task is to learn what to do in order to turn such challenges into not only survival but into greater accomplishment and a more full successful living experience not only for themselves but for others as well.

3.6 Government

We have to much government interference in our lives today. At ever level in the United States we have just far to many employed by government, to much taken in taxes by government, and to little done for our pursuit of life, liberty, and happiness. The federal government has a budget of over 4 trillion dollars now. Add in the 50 states, counties, cities, school districts, and other taxing authorities the entire "take" must be near 40% of a middle class working individual or family. A terrible tax load for every working person.

Also, government employees are paid very well now, with benefits included beyond the means that non-government works receive. Elected officials and appointed officials seem to do the best, with members of our Federal Congress at the top of the public trough.

In order to survive, changes must occur in all levels of government. Such changes must start in our basic written documents that are the foundation of our way of freedom, life, liberty, and happiness.

It can be and must be done!

3.7 Politics and Political Correctness

Politics is the art of compromise between opposing views on any subject of disagreement between interested parties, producing an acceptable outcome to all involved. In simple terms, an agreement that works for the good of all members of society. That is our hope!

Today across this world of ours, and the only one we humans now inhabit, politics has failed. That is the usual state throughout our history. However with globalization mankind has become totally interdependent on working agreeable relationships at all levels. Ease of movement of people and goods, communications, and Internet and the Webb have profoundly effected politics forever.

In order to survive politics must be come totally reliable between all levels of interest groups, especially regions of this world of ours. It is a challenge never faced before by our human race. Demigods cannot no longer be acceptable in positions of power, i.e. not in governmental, business, religious, financial, or military positions of great authority. If allowed, great power means great wrongs shall occur by such people that cannot ever be corrected.

People are easily deceived by broad smiles, vague promises, mirrored illusions, faceless utterances, glitter, applications of "snake oil', slick issued 'fact sheets', and professional run campaigns by well paid professionals. Lobbyists for big (very big) corporations, big unions, and well off (very rich) individuals, promote with vast sums of money extended elections cycles for their own benefit, not yours or the general populace.

They care less for your welfare, just there own. The State Legislatures and most certainly our Federal Congress is filled with millionaires and lawyers, living very well off of our tax dollars.

Mature level headed persons with common sense and proper judgement must be selected to occupy such positions at all levels. We still have a minority of same in some political positions, thank God.

To get such may people elected and appointed may be as impossible as waiting for pigs to fly! So what is the guide to survival in such a case.

Just keep trying to correct our organizations; religious, political, educational, business, etc ... and never ever give up on the goal. And get rid of life long professional politicians, limit terms, limit salaries, limit lawyers serving, etc ... by a alert educated involved populace.

And vote. Always vote. And try to vote for honest concerned individuals (if present on the ballot) that shall help improve our system of governments at all levels. As a suggestion, do not vote for anyone who espouses the madness of "political correctness" which is and will continue to wreck the structure of our society and country, as long as it is embraced by the radical few.

Much of "political correctness" has been formed into laws over the past 40 – 50 years, First, by the federal government, and then by many state governments. It was meant to correct society's perceived errors against women, people of color, people of LGBT characteristics, etc., that majority members of our society may commit. That is, to change most members of our society's outlooks, i.e. to change (correct) same, as perceived by some organizations or persons, by law, to their views.

For example, a "post man" is now a "post person" or a "secretary" is now a "office assistant" or "marriage" is no longer between a man and a woman, but between "two (or more?) people, regardless of sex". The foregoing is just simple examples, and the effects are much more extensive, with far more "political correctness" formed into law, enforced by government agencies, embraced by big business management, and most politicians, upon our entire country.

This kind of nonsense has been embraced by big business, the military, many religions, and other organizations, regardless of clear biblical passages in both old and new testaments, and historical lessons of prior societies. We all are forced to "toe the line", in this supposed "land of the free", regardless of our personnel beliefs, practical experience, and while pursuing our life's goals. Even God's ten comments are to be "re-interpreted" and made to "fit" political correctness needs.

We must now all be "political correct" at work, home, worship, and help enforce same into our family structures. At major (and minor) universities degrees are now offered in "Black Studies", "African-American Studies", "Women's Studies", "Family Structures", etc. to any "serious" student who wishes to improve our society's outlook, status, and growth, toward a better tomorrow. Even small business must serve its goals, regardless of the owners beliefs, as well as all educational organizations, all levels of governments, and all military branches (including combat units). Navy Seals, Army Ranges, Marines, etc …, with no exceptions, are included. What absolute total nonsense!!

Woe to any person's career, job, or work who even mentions that a LGBT, minority, person of color, or woman, was not qualified for management, or did not have the technical training required for the job. As the Queen of Hearts said in <u>Alice in Wonderland</u> "Off with his head!"

"Political Correctness" must be stopped. It has gone on too long. It is an element now in ruining our society. What is the readers opinion?

Also, part of the "political correctness" movement is the madness of big business subsidies, which culminated in the "too big to fail" gifts made by the federal congress under Presidents George W. Bush and Barrack Hussein Obama (both among the most destructive presidents this country has placed in that high office) to big business. Together, it s costs are still being totaled up, but probably shall run well over 2 –3 trillion dollars, if and when honest accounts are found.

When the failure of the "Great Society" act, signed by President Lyndon Johnson, and the "War on Drugs" acts, sponsored in parts by several of his successors, costs are included, it is no wonder that we tax payers have a federal debt now nearing 18 trillion dollars.

Such costs and total debt staggers a common person's senses, and cannot even be correctly comprehended, much less the now occurring repercussions understood and evaluated.

"Political Correctness" must itself be sent to the trash heap, and replaced by solid common sense and judgement, plus lots of hard work, performed by all, for a very long time to come to "right our ship of state". Pray that God Himself will give all of us the strength to do so.

3.8 Congress

The United States Senate and House of Representatives has not done its job for the people of the country at all well over the past "X" years. Together called "Congress", or "Congressional Members", instead of doing its job they were elected to do, it has benefited "Corporate America", and of course themselves. Together they have broken the great American middle class, the working class, as well as the "American Dream" of upward mobility.

There are no excuses at all for this remarkable lack of duty, elected by a free people at the polls, and pampered by all known laws of the land. Congressional members have made a lifetime of extortion of their citizens a great game enhancing themselves with vast lifetime of rewards, which the reader of this survival guide can never completely list, even with professional help from any group such as the nation's press.

The value of "X" may be filled in by the reader's experiences, but I, Omar, suggest at least 55, or back to the year 1960. Congressional stupidity or lack of common sense has resulted in wasting our military, industrial, financial, educational, and even spiritual capitol in tremendously very stupid creation of laws. This has completely burdened the middle class and working people of every level in our wonderful country with untold costs yet to be paid or even calculated.

In short, to correct this, some simple, but profound changes, are needed to our very basic documents of governance. This must be done for our way of life's survival, and soon!

3.9 Basic Freedoms

Our basic freedoms are at risk. Telephones are tapped, conversations recorded, Internet use is monitored, drones with cameras fly over our cities, data masses in computer data banks are "mined", our tax returns are opened by politicians, etc ... What has happened in the past 30 or more years to our basic freedoms? Cameras are mounted in all shopping centers, on police cars, and found now in all I pods, I tablets, desktop computers, mounted upon traffic lights, etc ... In short, cameras are almost everywhere, even in restrooms! Nothing is private, or able to be kept private by desire, wishes, or need of the individual. This must be changed!

Where is the freedom to keep our lives ours?

3.10 Conflicts & War

Conflicts are going to always be present with humans. It is part of the living landscape we all occupy. However, full-scale war on a large regional, much less world basis, cannot be allowed due to mass destructive weapons. Not just atomic, but biological, and chemical weapons are available that can cause vast damage to people and environment.

This is a incontestable fact in our 21st century. Period!

Total war has become to costly in human and material terms, plus collateral damage too extensive, and resulting suffering to great for it to be tolerated. Dictators around the world need to be removed by the world community with minimum suffering of their subjective peoples.

This is a United Nations responsibility that has been skirted up to this time. A force similar to NATO's military force should be on call, starting immediately, and be activated
and placed in continuous action against a agreed upon list of nations causing such troubles.

Russia is a good place to start, as no Hitler's "Libber Strum" (Living Room) type nonsense needs to be tolerated by nations ruled by a rogue person, or a group of rogue persons. Period!

This most certainly includes any aspect of terrorism, especially that supposed based upon any religious grounds. A persons religious views are a personal choice, and should never be forced upon anyone that does not freely desire same.

Common sense in doing so is easily is illustrated by the present terrorists activities in the mid-east (Arab nations), and Putin's actions against the Ukraine territories.

The ISIS terrorists in the mid-east and other terrorists organizations in that region should have been stopped long ago by the combined United Nations.

The United Nations organization needs a drastic overhaul in structure, operation, and function now, not later.

3.11 Avoid Fruitless Wars

At all costs the powerful nations of our world must learn to completely avoid "fruitless wars". These are defined as completely unnecessary conflicts between nations, which are caused by politicians who have "no skin in the game", or who think with their hormones instead of their brain, or wanting to "grind an ax" for some personal gain, knowing they themselves will not pay in any way for such a conflict.

Recent examples of "fruitless wars" are abundant, and include the following:

1.) Vietnam War – from 1963 to 1975, costing at least a total of a trillion dollars, at least 55,000 American lives, veterans (numbers unknown) requiring VA services for their remaining entire life span, acceptance of at least 300,000 Vietnamese refugees, etc. End result was a communist reunification of all Vietnam, a slaughter of at least 2 million Cambodians under a brutal communist group, and finally Laos under total communist control. All of former French Indo-China is still paying the price of this "fruitless war".

2.) Iraq War – from 2003 to 2011, costing at least a total of a trillion dollars, at least 6,000 American lives, veterans (again numbers unknown) requiring VA services for their remaining entire life span, acceptance of an unknown number of Iraq refugees (still coming), cost of rebuilding a "democratic" nation still going, rebuilding a government that will never function due to a tri-split ethnic country, etc. A total avoidable mess, but one that our national leadership said was necessary due to "weapons of mass destruction" that Iraq never possessed nor were never found. This war also created a new more dangerous enemy called ISL, which we now face in the same country and elsewhere.

3.) Afghanistan War – from 2001 to 2015, and has not yet ended, costing at least a total of 4,600 American lives, veterans (again numbers not accurately known) requiring VA services for their remaining entire live spans, acceptance of an unknown number of Afghanistan refugees (still coming), cost of trying to rebuild a nation (of sorts, but very tribal in nature), and this time involving NATO countries to some extent. This "war" has not been terminated either, but still rages at times primarily due to its neighbor Pakistan harboring the Talban and other Islamic terror groups. This war has spread into many areas of Pakistan due to the Pastune tribal influence, and perhaps even Saudi Arabia's influence.

The above are the major recent "fruitless wars" our country has very mistakenly gotten into, and even if our major objectives were accomplished in a short time, we over calculated, and stayed, to "rebuild", "help improve", and "give them democracy", etc. All of which the groups of power centers left clearly did not want. They just wanted us to get the hell out.

Left out, but should be included, in a lesser list of our cost in our human lives, money, etc. is the following:

1.) Nicaragua
2.) Colombia
3.) Chile
4.) Venezuela
5.) Somalia
6.) Lebanon
7.) Syria

There are other countries of course, in which we have been involved, but I leave the expansion of such a list to the reader. I will say for now that only Chile and Columbia seem to have worth the total costs this country, others, and the host countries have paid for results achieved.

"Fruitless Wars" are to be avoided at all costs.

I paraphrase General Douglas MacAuthor, and other good commanders: "Wars are to be waged quickly, decisively, resulting in completely defeated enemy, at a minimum cost, achieving a well predefined set of objectives. No other result is to be accepted."

"Then leave."

"Above all, remember, that any other result will be a moral comprise with evil."

"Fruitless Wars" are useless very expensive nonsense, and our country has been evolved in to many such efforts since the end of World War II. Our leadership. statesmanship, and other forces (we have allowed) have driven our society into such a current state of affairs.

This must change. And soon.

3.12 Family

The center piece of organized civil society is the family.

Yours, mine, and all families together, form customs, follow mores, and social rules that all in that selected culture are expected to adhere too and follow for acceptance. The family is the unit that is expected to have and raise replacements (our children), provide necessary fundamentals to not only stay alive but for its members to thrive (shelter, food, etc …), and to educate (schools, learning, common sense, judgement, etc …) and provide support (aid, protection, safety) as needed through a family members entire life.

Hopefully, also love, affection, morals, and all other very basic human emotions are fully developed in a individuals family experiences.

Finally, a very basic knowledge of God and a very correct philosophy of life's purpose is passed to each family member as he or she matures with age.

Today the family structure is under great stress and attack around this entire globe we inhabit. Many different country's governments, leaders, and power people have let their egos get out of control. Whatever the reasons
(there are many overlapping ones) are the family structure is under sedge by wars, poverty, lack of opportunity, misapplied religious views (radical Islam), etc …

This must be stopped.

3.13 Women

Women are the basis of all. Civilizations would never have occurred unless women could tame men (somewhat), create a "home" atmosphere (cook, have children, etc …) and in short make "home" a desired comfortable, safe (reasonable), and even perhaps sought companionship by men.

Thus, society starts long ago by such luring of men into "marriage" and a organized family structure.

Women are actually the stronger sex of the human race, starting with their "XX" chromosome cellular structure. They normally have a much higher pain ratio than men, seek to solve social problems with comprise rather than violence, exhibit more basic emotions, are endowed with a more carrying attitude, handle emergencies in a calmer fashion, and even live longer on less. Some times, even today, much less, and even in a servitude basis due to customs and/or poverty.

All women should be fully educated to their desired limits, be enable to vote, run for public office, work at any profession or trade, paid at a commencement level as men for the same work done, and have all rights as men under the laws of the land.

Women also have a greater responsibility than men, as they set the morals of the society they live in. And they conceive children, not men. This puts a greater and perhaps unfair burden upon women in their child bearing years. Nevertheless, it women, not men, that have that natural burden, and again, not men. But today a women need not have a child unless they want it, due to medical and drug advances that are available in most countries.

Nature gave men a physical need for women that is hard to, near impossible to, ignore or control, and women must always be on guard. They must remember they control men, not the opposite, unless they live in backward societies that have leadership employing cultural mores that are completely outmoded in our 21st century.

This is all part of survival, for all to be aware of, and to apply common sense, when dealing with women. All women should be aware of men's drive and act responsibly with same.

All women need to understand the greater opportunities they have through life in our current 21st century, more than any set of societies has ever provided in history to women.

May God bless women, May all men respect and treasure them.

3.14 Feminine Emphasis

Unfortunately our federal government, and some state governments, have, in the opinion of this author, gone overboard by placing by law and into practice, a "feminine emphasis". This means that our educational emphasis, business promotions, military rank advancements, etc ... now favors women, whenever they compete with men, for the same opportunities, the playing field is now tilted toward favoring women, instead of becoming equalized. It starts in the earliest grades of public schools, and becomes very evident at ever college and university across the United States as enrollments at such institutions have become over 60% women. A complete reversal of just 40 years ago.

Indeed, some female "activists" who have a male child even dress him in girls clothing, and encourage their male child to act as a girl. This creates a tremendous strain in their identity issues as they grow in age and maturity. However, such female "activists" may even reject men in their life's except to visit a "sperm bank" to become impregnated.

Indeed, today some 20% of white women and some 80% of black women have their babies without the benefits of marriage to a male husband. Again, government programs bear a large part of responsibility, and of course the individual women bear the ultimate burden, while the tax payer at all levels must help pay the unmarried woman's expenses with her fatherless children. Welfare is always necessary, and will all ways be so, but when self inflicted by a women upon herself (and then upon others) needs to be corrected.

Children require a stable secure environment to be provided from day one upon this earth.
Families were formed by societies to provide same, with both a father and mother in a as history clearly shows. Other wise, families collapse, societies follow, with governments failing, and finally, countries cease to exist.

Is this were we as a advanced nation are heading?

3.15 Men

Men are usually stronger, larger boned, having more muscle mass, and due to the testosterone hormone much more aggressive than women. They also have a "XY" chromosome cellular structure and a lower pain ratio than most women. Men cannot control anger as well as women, and compete for the "alpha" male position in any collection of people forming any kind of group interests. Therefore men always try to dominate each other in some fashion or another.

This causes conflicts that range from family abuses up to and including wars. It makes one wonder if man is really made in God's image, doesn't it?

However, men are made to "push the envelope", progress technically, improve the environment for a better level of living for all, to be inquisitive, be inventive, to innovate, to provide the means of living, and unfortunately, last but not least, to provide security, stability, safety, and protection to their chosen mate (wife), and offspring (if any).

However, one thing to remember is that men need women, and until nature is completely changed by God (no chance), always will. In other words, women can do without men much easier, than men without women.

Women stir deep, very deep, emotions within a man, even down into his very God given soul. This make a man in the state of "love" the most wonderful of creature, for his woman, or a most dangerous one, in a few unfortunate cases.

Men seem to run society, but are really run by their women folk (under cover of course). Usually this is a hidden truth, but not always. However some very backward societies, even today, especially in what the world geographers call southwest Asia, are well behind the 21st century, and physically (with religious overtones) are living some thousand years or so behind in a tribal state. Their present society sees their women as "objects" or "possessions" to be bartered, ordered about, and "dominated" in almost every sense by their alpha males.

Examples of very tribal states are as follows: Iran, Iraq, Saudi Arabia, Yemen, Somalia, Sudan, Rwanda, Nigeria, Israel, etc … Most are located in the middles east and in north Africa. Such nations found in those geographical locations usually have Islam as their major religion, have one race present, or a dominate ethnicity, or a combination of both. Plus a common culture that completely dominates any other that may be present, with women held in low esteem.

Most men of this world who fall into the "tribal" format are now doomed to a lesser life that the modern 21st century offers. A short, perhaps violent, wasted existence, being lead by their alpha males into a empty existence, here on SOL3, is usually the common result.

Again, this is part of survival, for all to be aware of, and to apply common sense, in dealing with men.

3.16 Children

Children are the God given product of marriage that continue our human race forward. They should be considered a divine product by not only their parents but by society. They are our real treasures.

Children have a right to a father and a mother as they are raised, nurtured, and schooled. Also, children have a right that both parents be adults (at least 18 years of age or older). Other rights children have are safety, security, and stability of environment (home, school, church).

Children have a right to proper nutritious food and drink, clean air, clean water, and clean clothes. Also, they all have a right to an education, as far as their abilities will carry them.

Today the United States, and much of the rest of the word, have the healthiest children and young adults ever seen on SOL3. Good sanitation has been a large part of this advancement, plus vaccinations, child labor laws, and criminal convictions of pedophiles.

For proper security of children, it is many people's opinion that pedophiles should be either put to death, or jailed at hard labor for their entire lives.

One last comment on children. It is a medical proven fact that parents should not be related, due to genetic inheritance of genes that cause mental or physical disabilities. Some societies even in this 21st century still condone, plan, or even demand such, and this often dooms the offspring to a lesser life.

Again, this is part of survival, for all to be aware of, and to apply common sense, in dealing with having children.

3.17 Marriage

Marriage in all religions that the author knows of, is between a man and a woman only, called monogamy. Other forms called "marriage" are polygamy, child marriage, temporary marriage, plural marriage, forced marriage, etc ...

Co-habitation is an agreement by interested parties to live together for common welfare.

That is quite different from that of a blessed sanctified marriage. No re-interpretation of the Holy Bible, Koran, Torah, etc ... as found in major religions, can refute this difference. None. Those that try are misleading themselves and their followers, in fact are destroying one of the foundations of society all over the world. Period!

Ruling by courts of any type, or convenient erroneous interpretations of history or selected religious passages can change what marriage is and must always remain.

Common sense, family security, societies stability, and even nature requires monogamy as marriage.

3.18 Friends

True lifetime friends are very few and far between. They are as rare as diamonds falling from the sky. A person can spend a lifetime of 70 years or much more, and be able to count true lifetime friends on the fingers of one hand.

This is a absolute fact, so security seekers remember this. Real friends are few and far between.

However, "networking" by creating associates interested in the same field, efforts, topic, interest, or other endeavors is highly recommended. Such associates are very necessary in today's commercial and/or social environments, and should be immensely helpful when needed. It's an application of the old "You scratch my back, and I will scratch yours" saying in a very practical way.

Such people may not be family, or even personal friends, but if they are trustworthy and reliable, it's a win-win situation for all concerned. Common sense with survival and security in mind is improved by networking.

Remember who you know is important.

3.19 Society

Mankind in various areas around this wonderful planet of ours have developed societies that vary in customs, mores, outlooks, relationships, beliefs, religions, and other interesting differences, with most differences due to terrain, availability of needed resources, weather, conflicts, trade, and even leadership of alpha men and women.

These many shapes and forms of societies, such as dictatorships, monarchies, socialism, communism, religious, communal, tribal, democratic, etc ... are too many times, no matter what the form, have religion and secular laws mixed, with a prime example being Islam's Sharia laws. Freedoms then vanish, and men seeking power preach it is God's will.

Of course, these pronouncements benefit the group seeking power. Never the people themselves.

It is the same with empires through history.
Empires have risen, dominated, and then fallen to more powerful forces. The Yuan and Gung dynasties, Persian Empire, Roman Empire, Inca Empire are just a few examples. Recently, in the 20th century, we have endured the Imperial Japanese Empire, The USSR Empire, and for a brief time, the German 3ed Reich under Hitler.

The most recent fall of a empire has been the USSR communist form under Russia's lead and dictatorship. This change, in my opinion, has benefited all of Europe, Russia and its people, and the entire world.

At the present time, in this second decade of the third millennia, societies all over the world look to the United States as the true bastion of democracy with true freedoms in written forms guarantee by its government that follows laws to implement the same.

Freedom, in all its forms and responsibilities, helps to provide not only safety, but security, and the individual's pursuit of life, liberties, and happiness on all levels for all peoples.

May God always bless America, and all its freedoms! May the entire planet's societies have same!

3.20 Safety

Safety should be provided by the society in which we live, and its government (at all levels) we live under.

However, safety always starts with the individual, then spouse, family, neighborhood, city, county, state, and finally country we live in. It also depends upon the individual's physical capabilities, weapon knowledge, weapon availability, common sense, judgement, and many other tangible as well as intangible assets.

Safety is a relative untouchable "thing" that cannot ever be absolutely guarantied to any individual, even to the President of the United States (remember JFK?).

All sorts of government agencies at every level from local police to the US Army have been created by society to provide various levels of safety. None are absolute.

Again, this is a absolute fact, so security seekers must remember this as well.

3.21 Environment

All aspects of the environment we inhabit are vital to a healthy productive well lived life.

They are hopefully all covered from a survival aspect in the following sections.

Beware of radical environmentalists, as they like any extremist, just lose site of common sense and sensible required tradeoffs. As an example is a fish found only in a ten mile stretch in the Sacramento River in California called "snail darter". The state government has mandated water be diverted from almond groves that need same to keep the water level required by the fish. Result is uprooting of almond trees, closing of many of orchards due to water allotment loss, jobs lost, etc. The fish will be lost anyway from the river environment in a few years.

Nature can be a very harsh master, with hard lessons, as our environment continually changes, and people have so very little control, as evolution still has the upper hand over all. Of course no right minded person wants to cause damage to our environment in this day and time. We need to repair past man caused natural environmental damages, and limit current such actions. Congressional laws are now in place with more needed to force those that cause environmental damage to repair same, with fines paid to enforce same.

Note what British Petroleum did in our Gulf of Mexico with their oil platform miscues.

Terrible environmental damage some of which cannot be corrected in many years.

Common sense is again required for environmental security now and for future generations to come.

3.22 Air

Clean air is a must for all to breath. Products that pollute our air and thus our lungs should not be allowed.

We all need clean air through out our entire life.

As an example, the state of California has done a wonderful job of cleaning their air. Los Angles air stung your eyes and the Hollywood sign across the valley could not be seen just
40 or so years ago. Enforced laws on vehicle exhausts, scrubbers in power plant stacks, burn bands, improved vehicle engines, etc … all contributed to achieving cleaner air.

Another example is the major cities in the Peoples Republic of China, where today they have such air pollution due to not having and enforcing clean air regulations entire populations that must wear masks or some type of breathing apparatus when outside. Children have days were their mothers are advised to keep them inside due to such poor air quality.

Air in our homes, offices, factories, etc … and air outside must be keep clean by all means possible.

This guide to survival strongly suggests a no smoking policy of any substances, and filtering your environmental air at home, work, and in your transportation choices.

Clean air is an absolute must.

3.23 Water

Clean water is a must for all to drink, bath in, and many other purposes. Products that pollute our water and our thus our environment should not be allowed to do so. We all need clean water through out our entire life just to live.

Polluted water means a very unhealthy environment in which to live. Cholera, dysentery, parasites, and many other diseases thrive in "dirty" water. Absence of proper water treatment plants producing "safe" water for use, and then after its use improper sanitation treatment of used water causes tremendous damage to people, environments, and untold costs to try to correct later.

For example, the state of California has done a wonderful job of providing water for their needs and usage. For example, Los Angles uses water brought several hundred miles by the California aqueduct system from northern parts of the state. Safe water has been provided to millions of people and farmlands (California's famous valleys for growing all sorts of needed food). That in a semi-desert area found in the entire southern half of the state.

Another example is the major cities in the Peoples Republic of China, where today they have terrible water pollution due to not having purification plants, and not enforcing clean water regulations upon local manufacturing concerns. This has put entire populations at risk. Dumping heavy metal residues from manufacturing plants, such as mercury, lead, and iron into rivers, lakes, or other water resources for disposal before proper treatment not only damages the entire environment, but at times kills people. This is very bad in the short and long time scales for any society.

To be fair, a few years back the Detroit River caught on fire because of "dumping" heavy chemicals. Also, areas of the Great Lakes have been polluted with industrial and city wastes not properly treated. Indeed, a great "dead" area exists in the Gulf of Mexico starting from the

Mississippi River delta area into the Gulf of Mexico. A very large area has been oxygen deprived from upper river wastes (herbicides, pesticides, fertilizers, etc ...) where fish can no longer survive at different times of the year.

Note that undeveloped countries, well over 120 out of 194 as members of the United Nations, have little or nonexistent water treatment for vast numbers of their population. Ditto for proper waste treatment and disposal. Humanity in the 21st century has a very long way to go in providing safe fresh water for our needs.

Safe fresh water is a necessity of life. Just as air is.

3.24 Food

Modern agriculture techniques using improved seeds of all grain stocks have increased yields all over the world. Proper use of fertilizers, herbicides, and pesticides together with crop rotation techniques and soil conservation methods have added to yields of corn, wheat, rice, soybeans, etc. … Mankind is now producing enough food for all current inhabitants of SOL3 for the first time in history.

In addition better storage techniques, transportation vehicles for food products (such as refrigeration trucks, train cars, and ships) are enabling better potential distribution of foodstuffs where and when needed.

The real problem is man himself, with local wars, politics, terrorist groups, black-markets, controlling food consortium's, etc … This has become the real problem in people not have the food, fiber, and drink that they need.

Such acts must be dealt with from a international level of concerned governments.

In the United States there are hungry people living below the poverty level for a series of reasons, but these are usually social and not supply or available reasons.

In any culture people of all ages should have proper food and drink for a healthy physical body. Growing children must have proper diets, and millions do not today around our world. What shame it is on governments, corporations in the food business, land usage, distribution systems, and religious leaders not doing what should be done.

Population control with proper sex attitudes is a large part of the problem today. Women are not just reproduction factories.

Being able to eat and eat properly is most certainly a part of security for most of us.

3.25 Housing

All people of this earth should have proper shelter, with necessary facilities for good health and welfare.

It is within reach this century, I think. The organization "Habitat for Humanity" sponsored by (among others) Jimmy Carter has certainly being showing the world a way to do so.

Perhaps people fighting "lost" causes will take notice and turn to a productive effort building homes.

Enough said on this topic, as a billion people today have no proper homes structures to live within.

3.26 Energy

The world is using so much energy that the demand cannot be met without constant power being added to the available grids of power lines and addition of oil and gas pipe lines.

New power generation methods (wind turbines, wave-motion turbines, solar panel farms, gas fired power plants, hydro-electric generators, etc ...) are being added in countries
around the world. Ships for oil and compressed natural gas are being built larger and larger to transport from such fields to user ports.

Even with energy efficient appliances, energy efficient light bulbs, automatic shutoff devices, etc ... more electrical energy is demanded and needed by users.

Commercial companies, governmental organizations, military branches, and individual power consumers are making great strides in conservation of electrical energy. As an example is the "sleep mode" not only available on computer equipment, but also being built into other appliances as well used in our homes to conserve electrical power.

To survive requires using all our resources in the most efficient manner we all can.

Energy is the basis of living standards.

3.27 Transportation

A good part of globalization is due to modern transportation systems abilities to move goods and people in mass like never seen throughout history.

Cruise ships that can carry thousands of people in comfort for weeks, tankers that can transport millions of gallons of oil, gasoline, or other liquid products, liquid natural gas ships that can carry compressed natural gas in huge quantities ply all our oceans at all seasons of the year. Airplanes that can be configured to carry up to 600 passengers are being flown now all over the world at all hours by airlines servicing many countries. Feeder airlines are abundant for national and international airlines. Trains consisting of 150 cars with capacities of tons pulled by diesel engines of tremendous power are common bulk carries across country sides all over the world.

People movers (moving sidewalks, escalators, elevators), high speed passenger trains, subway systems, etc … are readily available. Of course all kids of trucks, cars, motorbikes, bicycles, yachts, pleasure boats also are available for those that can afford same.

In other words, all peoples around SOL3 can move about with speed, and usually with safety in a secure fashion.

Transportation is not what it used to be, and one's survival is most certainly dependent upon this fact.

3.28 Entertainment and Communications

Movies, TV, Internet social software (Face Book, Twitter, Skype, etc ...), cellular availability (I pods, I tablets, cell phones of any type, etc ...), DVD's, satellite broadcasting, and even old fashion radio have changed the face of entertainment. Any medium produced in any country can be almost instantly be transmitted and dissimulated to people around the world as it happens.

Hardware and software are increasing this ability according to Morse's Law with greater speeds, tremendous reliability, and better definition as time passes each year or so.

In other words, all peoples around our SOL3 can now see, hear, and even discourse with greater rapidity interchanging all kinds of data, perhaps centered upon entertainment,(or business, or financial information) at unbelievable speeds about this world.

This is building a world community of common ordinary people, regardless of religion, customs, race, age, etc. ...

Again, for one to successful survive in such an entertainment rich environment, one must be adaptable, flexible, alert, and response with common sense. A constant ability to learn through his or her lifetime is certainly a basis for the applying the foregoing attributes.

Entertainment avenues broadens one's experiences, opening new thoughts, and drives changes as new communication tools become available to the individual.

Both together are changing the world's societies. We are not so different as human beings as in the past.

3.29 Sports

Our race needs a valid substitute for war and its resulting damages. Sports from the time of the ancient Greeks in the original Olympic contest 2500 years ago filled that need. It still does today in our 21st century.

All kinds of sports are competitive, either within one's self or with others.
A winner should be gracious to any opponent, and a looser accepts defeat in a similar manner (or should).

However, as the famous general, George S. Patton said, "America loves a winner, and hates a looser!".
That is human nature from a lifetime of observations.

A survivor must remember the statements in this paragraph, as sports is a "hot topic" subject at all levels. People always get overexcited and even very angry over trivial items such as sports.

Sometimes they even, believe it or not, kill each other over sports.

So a person that wants to survive, in all circumstances, remembers this.

3.30 Medicine

Medicine improvements and medical practice must be given great credit for its improvement's in very recent years, because it is the step-child of all sciences and engineering in the STEM areas.

No longer in this 21st century.

However, it nor its practitioners should be given all the credit for bettering our life's, increasing our life spans, or working miracles in repairing bodily injuries due to nature, infections, accidents, or a person's own stupidities.

Much more credit should be given to cleaner water, cleaner air, cleaner environment, improved housing, improved food stocks, implemented safety standards, garbage disposal methods, sewer systems, better transportation, etc. ... Probably these simple factors have been 90% of the improvements modern medicine is given credit for by our 21st society.

Remember it wasn't until around the year 1900 that most doctors before and after surgery washed and disinfected themselves and their operating rooms. Death in the United States was very usual for all men by age 45, and women perhaps a few years younger (childbirth was very hazardous at that time).

Now availability of hospitals, medical clinics, new drugs, new medical equipment's, emergency services, transportation vehicles, specialists in every conceivable part of the human anatomy, and finally team care concepts have improved medical care.

Also the same have made such industries fortunes, insurance companies unbelievable wealth, and the patient's bill payers broke.

However, picking a qualified trained experienced physician that has proven themselves able to perform correctly any (even simple) procedure

Omar

is really chance, as very few patients know much about their health care provider.

To survive, just ask. Do not be hesitant to do so, since its your life at risk.

Also remember that a medical treatment can either cure you, or in extreme cases kill you.

Investigate, look into, find out, inquire, compare, dig for common sense answers, etc ...

3.31 Drugs

Probably over 85% of drugs today at a pharmacy were not available just 15 years ago. Research has produced a vast array of pharmaceuticals for prescription use. Some can be very powerful upon the human body. In short, they can cure you, or they can kill you, sometimes very easily.

It really can get very confusing when a person is issued several prescriptions for a variety of illness by several specialists who are very good in their fields of medicine (heart, cancer, diabetes, bowl, etc …). A honest doctor, when asked, will tell you that even the medical community does not know the effects of many drugs given for several health problems a patient may have, when that patient is the victim of multiple illnesses at the same time.

A survivalist, that is that ill, must take extreme care, especially since most in such a state are older senior citizens. Society labels this period of life "retirement" or call it "the golden years" for some strange reason.

Finally, drug companies in the United States are overcharging for their products at every stage of care. They are getting all the traffic will bear, like Adam Smith stated in his <u>Wealth of Nations</u> book. Our wonderful politicians in our federal congress has warped health care to aid drug companies and insurance companies to charge the maximum.

To survive take prescribe drugs as the doctor or druggist directs.

Also, only take what is needed, no more.

Period!

3.32 Hospitals

Hospitals promise a great deal to those who must utilize their services.

However, promises are not a fact, until after they are totally kept. Costs are absolutely over the top! A hospital bill is a wonder in modern accounting. The language used is entirely from a foreign Latin manual, mixed with hidden costs unverified, and stated in a

mystic fashion never intended to be unraveled. Even by employing a numerous body of lawyers, accounts, cost analysts, and insurance specialists (including medical doctors working for health firms) your hospital bill will remain not understood by you and even those just listed. Believe me, it will.

As Sir Weston Churchill (the World War II leader of the former British Empire) once used the phrase " …it is an enigma wrapped in a riddle hidden in a mystery …", which today applies to any bill received by a survivor of such procedures. Assuming you do survive a hospital stay, of course.

The best course is never use any emergency hospital services (ER, its called), nor hopefully have any major surgery (for example a kidney transplant). If you do, prepare to be enslaved, perhaps for the rest of your natural life. You will never be able to pay the entire bill.

Your best bet to stay away from hospitals is to live a healthy lifestyle: no alcoholic drinks, no tobacco products, no illicit drugs, no wild living habits, proper exercise daily, and a very healthy diet always. In addition pray that your parents passed along very good genetics. If all of the above is ingrained into a very moderate life style, and in absence of major injuries you may be able to only be in a hospital twice in a your entire life: at birth and at death (perhaps).

This is the very best a survivor can do in this paragraph's topic.

Sorry!

3.33 Doctors

Doctors are usually at the top of the intelligence chain, well trained (you hope), and really practice medicine for the benefit of their patient, who they know well (not just met). Also, they keep up with the latest techniques in their field of expertise, and have control of their own egos, so if consultations with another of their profession is necessary they will do so at once, again for your benefit.

Doctors have been told (but you probably have not) that whatever ailment you have, if you live a healthy lifestyle that is reasonable (common sense used), approximately 85% of the time your body heals itself, approximately 10% of the time their issue of the proper drugs and treatment will cure your problem, and 5% of the time, no matter what they do, try, or recommend, will not help you at all.

The above is a fact from a fine doctor that I known for years. It is a fact in medical schools as well.

However, remember all doctors are not equal, and a second or even third opinion should be sought if really needed by you, the patient.

Also remember doctors are human. They make mistakes, just as we all do. They have great stresses at times just as we all experience during our give space-time life's here on SOL3. Also, they face a foe called death.

Death wins in the end, always has and always will, even for the best of survivors.

Remember that doctors are trying to defeat the undefeatable foe.

Never worry about it, just make the best of life. Live it to the fullest! God wants us all to do so!

Omar

Keep your health to enhance your living!

Just choose your doctors with care, and avoid those (very few) who really should not be in a medical profession.

3.34 Retirement/Recovery Centers

Retirement centers are places which our modern lifestyle has created to place our old parents or relatives in, until they pass out of this existence, to their heavenly reward. They fill a need in a fast paced advanced modern society that relieves people of actually spending time to care for the previous generations that are still breathing.

Such care used to be expected from the younger healthier people in each family just a few generations ago. Not now!

We off load such responsibilities, with money spent of course, onto a business of quote "care for the aged". This is our answer to a pressure packed constant on the go get it done globalizes work approach and living environment modern societies have created.

A retirement center has no family atmosphere, no children present, no teenagers growing up, no young adults moving on in life, no family adults present, etc … It is a "dumping ground" for old worn down people at the very end of life who have been removed from their home environments. Usually they contain mostly old women. They are then out of sight and out of mind from any family presence or worry.

Recover Centers are a bit different. They employ a variety of physical therapists, nurses, aids, etc and visiting doctors monitoring their patients progress. However, today they operate under a "for profit" motivation regiment that demands from insurance companies and government legal contracts to "get results quick".

Recovery Centers do have a place that is needed by many families. Their services can save a family caregiver from working themselves to a early death because of lack of proper training, equipment, needed skills, experience, or just plain strength with proper stamina.

However, special insurance is needed to cover Recovery Center costs, and is very expensive as you would expect. Also, most families do not recognize the need for it until to late.

This same set of "get results quick" policies is found in hospitals as well, with the end result of at times moving patients up and out to fast. These types of actions can cause terrible damage to the patient's recovery, and even cause death. Stroke treatment centers in hospitals are notorious for moving patients out to Recovery Centers before they should.

If at all possible, a aged person should die with dignity in their familiar home environment, as was the case a few generations back. It seems wrong to die surrounded by strangers in a strange setting. Proper good byes should be said and love expressed before each leaves for God's judgment.

Life unfortunately, is not always the way we all would like. Especially our last remaining years. In this 21st century most senior citizens will die in intensive care units (ICU) in hospitals with strangers present at the time of their passing.

As a survivor, what would you choose, if allowed?

3.35 Extended Life Span's

In the year 1900 the average life span of a man was approximately 45 years. By that age most men then died.

At that time women lived even less, due to childbirth complications. Ever wonder why our hospitals in the Unites States almost always have special wings for women and children? That's why. Men were tired of loosing their loved ones.

Today the average life span of the average man is approximately 79 years, and a average woman lives to 82.

Just what are we to do with these extra years added to our life spans? Have two working careers, or enter retirement much later than 65, or spend the extra years doing public service, or perhaps having extra "Golden Years" to just enjoy doing whatever you can afford?

Extra years could be spent by helping your children or grandchildren in their pursuits. Any number or combinations seem open to choice. What is certain that more older people with insufficient retirement funds will be present around the world. How will they make do in a financial sense? What extra strains will be placed upon retirement pools of money?

An open question is health care for such expected large numbers of older citizens. Also, if they are not in good physical shape, what quality of life remains for them, and their care givers?

It should be a law that all workers have a mandatory amount deducted from each paycheck that cannot be touched, and is invested, until official retirement is reached. Such an amount must be calculated, and set aside for the retirement years. It should change yearly to account for financial global changes.

Australia does this now.

For your own safety and security give this a lot of thought for old age survival. What will you do with extra years? Help kids in school? Aid your church? Be a volunteer for civic duties? Open a second or third career? Help the family business?

Again, just what will you do to help society with added years?

A warning: live a healthy life style throughout your entire lifetime to gain those extra viable useful years. Common sense it is.

Be a lifelong survivor, and a lifelong contributor!

3.36 Education

More, not less, education for every child, women, and man that inhibits this 21st century is required on a lifetime basis now, and in any foreseeable future, in every country on this planet.

It will have to be on a continuous basis, using every technological tool we have now or will develop, to deliver top quality education on topics delivered where and when needed to each individual, in a proper learning format that is understandable by the student.

Such educational materials must be affordable from pre-kindergarten through university graduate level to all students of any age, sex, religion, race, etc. ... No one should be denied education that they desire, need, or can use to make life better.

Internet, the Webb, DVD's, educational TV, etc ... have made a wonderful start, but lectures by master teachers must be made available on all subjects in reliable formats.

It must be remembered by all that any legal work endeavors (trade, profession, needed skill, etc ...) are all honorable to perform, and equal in need by society, now matter what monitory remuneration is paid for doing so. A bank president and a sanitation worker are just as vital to our operating society. No job is unimportant.

Plan on a continuous education effort occurring at various levels for your entire life.

This is needed for survival, believe me.

3.37 Trades

Joseph and Jesus Christ were carpenters, Joshua was a soldier, Mohammedan a merchant, and many others in holy texts practiced a trade for a living. Many of the Christian disciples were fishermen.

There were no "free" lunches or "free" rides in life for any of them.

Tradesmen make, repair, maintain, and install the needed essentials in modern 21st century societies. They are the basis, along with farmers, ranchers, mechanics, electronic technicians, programmers, plumbers, and many others to numerous to mention that keep our living standards going and improving. They are the backbone of society's body.

Internet, the Webb, computers, the "cloud", and every other automatic system you can think of are highly important and indispensable to the globalizes society of today, and they require all sorts of tradesman.

Bill Gates stated once that each professional person alive today should also master a trade.

In other words, get a set of skills that can be used as a tradesman, as well as a profession, IE … a computer scientist (professional) can also be a programmer (tradesman).

This means survival, big time.

3.38 Professions

Medicine, dental, law, politics, engineering, teaching, ministry, military, etc ... are thought of as the major professions one may formally aspire to enter, attain a formal education in, and spend a lifetime practicing same in the public "eye".

Accounting, banking, investment, insurance, financing, marketing, etc ... fall under the wide field of business, and may or may not be considered as a major profession, but one can also be formally educated and then practice its disciplines as well.

Professionals are expected to always keep current, perform to their professions best efforts at all times, and extend their fields of knowledge whenever possible while solving any presented problem set they encounter.

University night classes, weekend classes, seminars, and commercial seminars offered by corporations on subjects they are promoting are part of continuing education. Internet classes are available for self improvement, and even state unemployment centers my have classes of interest. The military services sponsor schools at ever level to help train their members in needed skills. Helicopter mechanics is just one example. Trade manuals and technical books on needed subject are available at libraries or from Amazon or other Internet online companies.

In any case, keep a home library, no matter how small, and keep up some form of self improvement in your trade(s) and professional skill(s).

It requires a lifetime of constant education either formal, informal, or self inspired to be equipped and competitive in this 21st century.

Again, this means survival, big time.

3.39 Colleges

A two year diploma in some trade such as welding, engine repair, aircraft maintenance, Internet servers, database upkeep, electronic technician, real-estate sales, programming, x-ray technician, teachers aid, nurse's aid, etc. … allows a trained person with such a diploma and possible certifications to make a good lifetime living.

Get a diploma with as many certificates as possible.

These types of jobs cannot be sent overseas by large companies taking advantage of globalization and available low labor rates in other countries.

Manufacturing jobs can be sent readily overseas, but service, repair, and support jobs cannot be, as well as construction jobs. They are here. It has been estimated that 80% of all trades and professions cannot be sent overseas.

As usual, the press makes it seem that 100% of all jobs can be done in a more cost effective manner by being sent out of the country, but this is not true at all. Surprised?

Many leaders of large corporations state that a person should know a trade skill, and also have a professional skill. This allows a better assurance that in any period of unemployment any part in the nation's economy that "turns down" can be more easily survivable by the affected employee caught in a layoff, plant closure, riff, etc …

Such a claim is open to much discussion. It would certainly help to have both if possible, I think.

Be flexible, adaptable, resilient, focused, and use common sense, always.

Survival during your lifetime is the end goal.

3.40 Universities

Universities in the United States lead the world's higher educational institutions in quality of educational, research, and innovation in all aspects of the terms. However, there is a tremendous drawback now in play at our higher educational institutions. It is called cost.

The deregulation mania that swept over the country approximately 20 years ago (approximately 1990 time frame) that produced corporations that own and run for profit charter schools, massive numbers of toll roads, for profit prisons, etc ... hit the university systems in almost every state like a ocean tsunami. Almost all state legislatures, our federal congress, banks, and other loan institutions rushed to set up laws and avenues for parents and students alike to borrow money for college and university costs. There was money to made in a new untapped market!

Today in 2014 a staggering educational debt of approximately 1.5 trillion dollars is owed by students, former students, and parents for attending universities, trades schools, commercial educational institutions, etc ... Some loans will take not just years but decades to pay off. It is another millstone to drag our future economy down, burden people who wanted a advanced education, a better life, must now delay home purchases, marriages, etc ...

But in our myopic vision of economies, its corporate capitalism at its best. This need not have happen, if deregulation of education had not been allowed by politicians in our state legislatures and federal congress.

And what have universities done with this new money source? They have exploded managerial staffs, increased sports activities, built new football stadiums, hired foreign instructors to "hold down costs" (many cannot be understood due to poor English languish skills), but not improved their delivery of a proper university education.

In other words, cheated their customers (students) by not improving their product (an education that is meaningful). Another method

Universities may use to cheat their students is the importation of foreign professors, many of which cannot speak clear understandable English, nor or qualified professional level teachers, but accept such positions at much lower salaries than normal.

If you decide to attend a university, choose same with extreme care, due to cost, necessary debt that most likely will come in the form of loans, missed wages during those years spent acquiring the desired knowledge, and finally the energy required to reach graduation.

You can cut costs dramatically by attending a local community college for the first two years of university required studies, while living at home. But make sure that credits earned are transferable to the major university of your choice first. Check with both institutions.

Do not count on working during those years of intense education, nor changing majors during those years. You cannot work and learn at the same time. Acquiring a proper professional education is a full time job. Know what you want to become in a professional sense before you start the effort. Do not waste money nor time. Neither can ever be made back.

Just remember, in order to survive a person needs a continuing quality education all his life in his trade and profession. This is now required in our 21st century.

3.41 Continuing Needs

As long as a person lives, all of the needs discussed in paragraphs of this section 3.0 will be continuing, and always needed to be taken care of by the individual, so at times, when

help is necessary, find it. Remember to establish a network of people: fellow workers, family, members of organizations you are enrolled in, etc … and do not fail to use any type of network members for aid when needed. You can always return the favors later.

Networking is extremely important.

Remember to prioritize your needs, and always keep after moral, legal, and acceptable ways to find solutions to satisfying your continuing needs.

Action is a must.

Common sense is a must.

Even prayer is a must.

Never quit finding a way to survive.

3.42 Travel Value

Travel broadens experiences needed for better judgment, provides much needed comparisons, allows new ideas to be generated for consideration, and improves one's information on various differences between customs and values of societies. In this era of total globalization, more people travel across boarders and around the world than ever before. Educational experiences have been enhanced by students attending foreign universities and friendships made between host and temporary residents that may last a lifetime. Comparisons of ordinary people's lifestyles can make lasting and very helpful impressions.

It may be that travel and temporary resident in a foreign environment will add to potential understanding of others, which hopefully results in a more peaceful world environment between governments as well. It is a established fact that more people travel between and then reside in for various periods of time in countries not of their own legally in this 21st century than ever before.

A better more cooperative and even peaceful world it should become as travel enables people to become aware of their common humanity and differences are put aside or understood.

In that sense globalization has been a good thing. It aids in survival.

3.43 The Near Future

In the near future (20 – 30 years) the world will globalize in a tighter fashion, because of corporation adjustments on their costs and profit factors. Communications will get faster, with unbelievable amounts of data being moved daily, the "clouds" of computers for storage, manipulation, and "mining" such data will be enormous in capability. The vast "network of everything" will electronically tie the entire global structure together. All people will have Internet access to the "Webb" for all sorts of communication uses (Skype is today's example, and its capabilities are just getting started).

Software applications (SW-PPS) will carry all of us into the deeper abstraction of levels of total computerization, and toward eventual control or overseeing (depending upon your view) almost all aspects of our daily lives.

STEM technologies, as basic theories continue to evolve, will produce more devices for enhancing our living than now visualized in this year of 2014.

Now over 2 billion people have cell phones in a world population of 7 billion. In the near future of 9 billion (2040) over 7 billion will have what then will be called their personal electronic extensions into the World Wide Instant Access (WWIA) Network.

Survival in a globalize environment is a must. Life as it will then be lived can be, will be, should be, extremely rewarding in every sense of the word.

The changes for the near future are already in place.

The survivor will use such changes to his or her benefit.

3.44 The Far Future

The far future (60-130 years) will have mankind (with women of course) on our moon, Mars, selected asteroids in the asteroid belt, and probably on other moons of Jupiter or Saturn as well, in established colonies and outposts. Such people will be carrying out terra farming, research, and commercial ventures yet to be thought of at this time. Chemical propulsion means for rocket engines will have been replaced by much more advance types of propulsion based upon a better understanding of quantum mechanics and advancement of theories with supporting mathematics allowing engineering of new types of propulsion devices. Perhaps basic physics will have found a way for travel faster than light in out space-time by using a now unknown dimensional bypass.

The normal life span will cover at least 100 years of activity productivity, and in the later parts of the far future probably a productivity period span closer to 125 years, due to organ replacement, medicines to control (delay) the aging process, diet improvements, periodic mandatory wellness checkups, vast improvements in the total environment of living, and finally a more civilized attitude toward life's value.

Life will be enhanced in the far future much more than experienced today. It will be longer, fuller, and healthier in physical, mental, emotional, spiritual aspects not yet seen nor even dreamed of today. Society will evolved to eliminate the age-old curses of war and poverty. In the far future mankind will fully realize that war and poverty are to be completely eliminated from all human societies.

This can and must be done with common sense, grit, and extreme determination to ensure basic freedoms along with life, liberty, and pursuit of happiness, is secure for all.

3.45 Technology

STEM technologies will soon have achieved a expressed mathematical "theory of everything" (T.O.E.), based upon a merger of the laws of physics in quantum mechanics and the laws of physics covering the large scale of the universe as found in Einstein's General Theory of Relativity. Together they explain the total foundations of our existing universe that God The Creator has placed us in for our space-time existence.

As the 20th century produced vast applications of electronic knowledge, the 21st is now and will continue vast applications of biological knowledge as the life science fields improve in basic research, then the 22ed century will produce vast space exploration opportunities.

For technology to continue progress, basic research in all STEM fields always must be emphasized, and STEM fields must include the medical and political areas. STEM must become STEMMP (Software, Technical, Engineering, Mathematics, Medical, and Politics).

STEMMP will make us a race that God Himself will be glad he created.

3.46 Internet & the Webb

Use's of the Internet and Webb have just really started, and as long as they are kept free of regulations that strangle and limit their offering faster, more, and innovative services, growth shall continue at an unbelievable pace.

Such growth of the Internet, Webb, "Cloud Centers", data mining, etc … shall then continue, increasing on a exponential curve of progress, and expansion of uses. In the far future Internet and the Webb will be on not only SOL3's moon, but Mars, the asteroid belt on occupied asteroid's, etc … This will tie human beings together regardless of their locations. The economic uses of this "system of everyone" will allow social interaction of various people's ideas and values to create a resulting mix of yet undreamed progress.

It is very feasible that in the far future almost all human interchange will occur over a almost immeasurable improved Internet and Webb like structure in a instant automatic fashion. It will also be completely fail proof, absolutely reliable at all times, self repairing, and totally secure.

This will be happening in stages already started in today's STEM environments as application techniques improve. However, like all human learning endeavors, it will take many years of great labor by many types of individuals and teams of same to achieve.

Security of the networks, called Internet for now, and perhaps in the future it will be called "Society's-Net", is paramount.

Proper freedom and use of "Society's-Net" is required for survival of all.

3.47 Computers

Quantum computers are now just starting into use. This represents the ENIAC stage that was first created for computing US Army gunnery tables in World War II, which first brought electronic computers into use in 1946. Look at were this development has led to electronic computer advances since then, with the added level of software instructions, under the guidance of software operating systems, with an astonishing array of software tools, all used for development of applications and controlled execution of same useful in almost an uncountable fashion.

Vast databases of stored information on all subjects have been created for a record number of uses. Financial, investment, historical, scientific, governmental, records, personal, medical, etc ... are available for multiple uses.

Using computers that contain derivatives of various electronic chips called processors that execute a set of instructions in a orderly logical fashion guided by programmed software has allowed tremendous advances in communications, control applications, data manipulation, medical applications, all fields of science, computer gaming, networking, etc ...

In the near future bio-electronic computers will be developed that perhaps mimic the human brain or come very close to doing so, with a vast increases in computational power, taking full advantage of mass parallel concepts, solving simultaneous parts of problems, with a series of solutions offered to instantly pick the best. In the far future, as more is learned in artificial intelligence areas perhaps computers will be able to advance themselves in capabilities for mankind's benefit.

It should bring all of us greater security in ever meaning of the term.

3.48 Quantum Mechanics

The engineering of computers and other devices using quantum based chips that utilize physics' current known quantum mechanics laws are now beginning to come into the application sphere. The physical underpinning of such advances depends upon how matter's basic sub-atomic particles act at the Plank level, That is, how matter acts as formed by basic vibrating energy strings. We are use to using materials down to the molecular level and no further, according to the laws of physics that Newton and Einstein's general and specific laws of relativity support.

Now quantum mechanics has brought us to the most basic level found possible, that of the realm of the subatomic level and how nature's most basic components act in the Plank length space-time frame. In physics, the Planck length is a unit of length, equal to 1.616199(97)×10 to -35 power, in meters (unbelievable small length). It is a base unit in the system of Planck units, that physicist Max Planck defined. The Planck length is defined from three fundamental physical constants: the speed of light in a vacuum, the Planck constant, and the gravitational constant.

At this level common sense as human being's know it seems to be turned upside down. It will take all of us a good deal of time to acknowledge and accept it as fact. It is suggested that the reader look for articles or a good book for references on Max Planck, the founder of Quantum Mechanics, along with any good book on Albert Einstein. See the appropriate appendix.

In fact, select several books written clearly by learned theoretical physicists, and then try to understand just what quantum mechanics implies. It will not be easy. Its not common sense, but very deep reasoning, at the most fundamental level of physics and mathematics.

It's necessary to have at least an overview understanding of Quantum Mechanics for complete survival (Omar believes) in the 21st century.

A true survivalist will build at least a minimum understanding of quantum mechanics.

3.49 Drones

Already there are several dozens of drones of various uses and types. Not just military, but commercial users have pushed drone technology into a world wide race for better designs, prototypes, and manufacturing cost effective techniques. Security uses by law enforcement agencies (police, boarder patrol, FBI, CIA, etc ...) and all levels of government (cities, counties, states) uses for monitoring or data collections purposes (highways, maintenance checking, traffic surveillance, building monitoring, school safety issues, property tax comparisons, etc ...) and many other applications by inexpensive cost effective drones with different kinds of sensors (cameras, infrared, speakers, relay transmitters, etc ...) and timers plus GPS location abilities are available.

Commercial uses are many with more capabilities being added every day. Examples are ranches monitoring and counting their livestock, farmers having drone attachments for crop dusting, and containing sensors for moisture needed for proper irrigation. Perhaps some day a drone attachment with a homing device will be offered for your automobile as a driver can have a look at traffic problems. Drones may deliver grocery packages to your doorstep, or drugs to isolated locations (Merck in Germany is already doing so).

Drones are usually thought of as those that fly, but there are also ground and water drones. Water borne drones (surface and underwater) are being developed to map the ocean, river, and lake bottoms. Also, they will be used in monitoring all sorts of parameters: ocean currents, mineral contents, energy generation, tracking of shipping, estimating fish stocks, etc ...

Defensive or offensive purposes by navies using surface and underwater drones will not be mention here. Neither will air force or army uses.

However, commercial uses will also include training pilots, delivery of purchased goods, monitoring children, entertainment, fire fighting, etc ... Drones will be valuable in ensuring your future safety and security.

Learn about drones, and drone "swarms", and drone intercommunication, etc ...

A survivalist will encounter drone usage in life very shortly, even if it's a new toy for his children.

3.50 Space & Its Uses

Close-space (100 to 22,300 miles from the earth's surface) has been filled with satellites placed in orbit for many different uses: communications, broadcasting, weather observations, earth mineral location, land surveillance, ocean surveillance, global positioning (a galaxy of 32 satellites), crop monitoring, vegetation coverage, logging, basic astrophysical research (the Hubble and Kepler telescopes), etc. ...

There are so many uses of satellites at different distances a complete list would take several pages. But laws governing the use of close space have yet to be agreed upon, and only some international agreements now exist, without a full set of laws as seen by all countries interested in governing its use.

Note that military uses and purposes are not mentioned at all in this topic discussion.

Mid-space (234,000 miles to our moon) has had probes sent to the moon for different purposes, and astronauts actually land on the moon for exploration purposes. Searches have been made by such activities seeking water, and mineral content determination of moon rocks. The mapping of the moon has been carried out by satellites constructed for that orbiting purpose.

In the future manned bases on the far side of the moon will no doubt be established for astronomical and basic research needed for future planetary exploration knowledge needed as man moves to occupy this Solar System we occupy. Commercial opportunities will have many opportunities using our moon. However, international space laws have yet to be put in place, and are needed to lay the basis for moon use.

Deep-space probes have been launched to Mars and other planets or their moons (Jovan system, Saturn system, etc ...) in order to gather better data for analysis and potential study and future use of same. No doubt that by 2050 man will visit Mars, and establish outposts with Terra farming in place as soon as possible. Today two "rovers" are exploring small parts of the planet by sampling soil and atmosphere with the ability of mapping some geological structures. Data is transmitted back to our receiving facilities arranged around our planet at selected locations and commands are transmitted to the "rovers" from the same facilities.

An individual's security in space will be at risk due to many unseen factors, and the inherent dangers of the environment. But all can be eventually overcome, as our God wills and man preservers.

Survival in space is and will be hard-fought to achieve.

3.51 Law, Lawyers, and Legal Entanglements

Laws must be enacted to cover such advances as are now occurring in such fields as: cloning, robotics, artificial intelligence, commercial space exploration, immigration rights, population control, crop seed hybrids, surrogate mothers, sperm donors, fresh water allowances, etc …

The courts must speed up in their decision making, and lawyers must be more efficient in handling clients. Today's massive backlogs in courts at all levels, and today's handling of cases by lawyers for creating maximum billing hours must cease.

Legal entanglements that we have today occur because lawyers are not specific enough in formulating laws. They try to cover all potential cases with to many generalities in their written laws, and add amendments that are not pertinent to solving the main legal question being address in the new law being constructed. This is a added burden for the courts to content with at all levels.

In addition. legal entanglements occur because courts, especially the Supreme Court, issue rulings that are not based upon common sense, prior rulings, or even the body of known law, but instead are based upon political leanings of the justices. Also, the courts are just to slow in hearing a case, as many lawyers get paid by the charged hour, and are allowed to "drag out" the processes on a case, on ever imagined technicality, no matter how absurd. In short, many judges are doing a very poor job of moving their case loads toward a final conclusion.

Finally, legal entanglements occur because of legislatures shaping laws in deliberate fashion to be argumentative in scope, when passed and applied, so as to create "work" (i.e., income) for lawyers, when contenting parties use or object to the new or amended statures. Law is the only known field were the major benefits that result go to the lawyers. Farmers produce food for others, doctors hopefully heal the ill and injured, teachers educate their students, police protect the populace

from criminals, etc … However, lawyers can produce a paper stream that can go on for years at tremendous costs to clients,
and have no results worthy of the time, effort, and treasure spent, except their own enrichment at the tax payers and /or their clients expense.

None are secure nor are safe in life without basic known legal guaranties with enforce of same. Today in our society and indeed the world this seems to be missing to a very large extent.

Some sort of legal society earthquake at the highest level imagined must occur to get lawyers and politicians off their current approaches and nonsense, stop them from just sucking society dry of funds, and actually do the above in a sensible manner.

It's a must for survival of our states, nation, region, and the entire world at large.

Law should lead, not follow societies needs. It should make common sense and be understandable by all of us.

Proper survival demands it.

3.52 Today's Needs

Greater security for all individuals on our SOL3 and between all nations and societies will be needed before such giant leaps forward can be made.

War, with its resulting destruction and harms, must come to an end, so peace dividends can be had by all, with appropriate population control, efficient resource use, including water, along with environments used in a maximum fashion, can be accomplished. In order to completely outlaw war the current United Nations must be either upgraded or replaced completely by a organization that is structured by eliminating the structured deficiencies it currently has shown.

A World Constitution, a World Bill of Absolute Rights for All People, and a World Bill of Police Powers must be very carefully drawn up by the best legal minds available, then made available with a time line to implement same. Such implementation would not take place until all regions of the earth's peoples agreed to live under the documents, with a United Nations Assembly (restructured as given in following paragraphs) governing bodies implementing the new world's constitution structure, the new peoples absolute rights, and enforcing same with the new world police.

First, all recognized nations must be members. No choice, but all nations must be a member. It is suggested that a lower governing body have a number of representatives, based upon population of their nation, and they are to be elected by the adult citizens of that nation in a free, open, and monitored manner.

Second, a upper governing body have a number of representatives, based upon the count of nations, and divided equally between all nations. This is very similar to the structure of the current congress of the United States. However, each body can introduce bills of any type, including revenue raising, or agencies to be created or eliminated as needed for various purposed, etc ..., for example, under the World

Police a temporary military command to enforce a set of laws, eradicate a terrorist group, etc … may be formed for a limited time to carry out the particular enforcement need.

The lower and upper governing body members may serve a period of only 7 years. Then they must resign and can never serve again in either body.

Third, a series of regional World Courts will be established, with judges form their respective regions to form a tribunal for each type of court (civil and criminal), and sit on such a court for a limited period of time, again say for 7 years, as an example. Then they will be replaced and can never occupy such a "bench" again.

No political parties will be allowed to form in either body, and the elected head of each (by the members majority vote) can serve as the moderator for only one year during his seven year term.

The military commands to enforce World Government laws and rulings are to be rotated between commandeers who must be totally responsible for following rules of strict engagement that have been given to them by the World Congress for settling the dispute at hand.

Other approaches can be recommended, studied, and all sorts of save guards can be inserted, but today the world deserves a world without war, famine, population explosion, religious extremism, terrorists, dictators, environmental damaging practices, etc … The world deserves sanitary waste disposal, clean air and clean fresh water, etc …

Mankind has come a very long way to date, but we can and must do better for our own safety and security.

A functioning overseeing world government has become necessary.

What is needed now, is population control, and elimination of religious extremism!

Asia (which includes India, Indonesia, Bangladesh, China, and Malaysia), and Africa (which includes Egypt, Nigeria and other examples) are already vastly overpopulated. Yet their societies, religions, and customs seem to prevent measures needed to at least start controlling population growth. Our earth cannot continue with such an expansion.

Also needed now is emigration control, from these overpopulated countries, as limits have been reached even by European and the United States. Governments must take responsibility, as well as religious leaders of all faiths, in this effort.

Elimination of religious extremism (terrorists) is ongoing now, but at a terrible cost. Just read the daily headlines.

Finally, war must go. Societies must see to it. It is not the answer. War only offers mass destruction of any society.

May God help us, and have mercy if we do not take action, country by country, region by region. Soon!

3.53 Tomorrow

The rest of the 21st century will have many challenges requiring very correct actions to met and solve them.

However, they all possibly could be solved. Especially those that ensure proper safety and security for people's rights and welfare of a free and open society for all.

A true world government shall be evolved from our current United Nations structures. World wars and regional conflicts shall be totally eliminated, as well as terrorism, with the eradication of same completely. Such acts will never again be allowed.

New laws needed will be strictly enforced, modification of governments, enforced environmental rules, financial system changes, population limitations, and above all religion tolerances, elevation of women's rights, and other now foreseen (and unforeseen) changes will be forced upon all nations and peoples driven by common sense, logical needs, and rational reason by societies to survive.

Vast technical advances shall be achieved with more underway, and mankind will not tolerate evil as we know it today in the second decade of the 21st century.

The first day of the 22nd century will be the dawn of coming unbelievable achievement for mankind, as long as we keep the basic freedoms open for all to have and to treasure.

The following quote taken from a famous James Joyce poem will ring with truth and clarity to all of our world's inhabitants, and those 'off world' (our Moon, Mars, etc.) as well: "No man is an island, No man stands alone".

In short, together the world survives, but divided the world has chaos.

4.0 Summary

These previous pages contain thoughts and views formulated by Omar. He claims to have the wisdom of 3000 years, or more, of living here on our God's planet we call SOL3 (Earth).

You, the reader, may think what you may of this unbelievable statement, but in any case Omar does shine thoughts on many vital topics that cannot be dismissed lightly.

In other words, its not all just bull stuff (no cuss words allowed). Just think about it. What topics would you add, change, or drop? Your safety and security concerns? Your solutions? Your actions to ensure family, community, and society obtaining a better world for all?

Many topics are not in Omar's list, but I am sure they will be added in the future to this start of his. However, Omar's thoughts on many subjects have been captured in this document's paragraphs. They are worth reading. Some may inspire you.

Omar's poetry now follows in section 5. Perhaps it will touch your heart and even your soul.

Please read it. Think about each poem, and what it means to your inner self.

Poetry can be from a writer's very inner essence, and can cause deep feelings in the reader, as its message is conveyed to the reader.

Poetry can be beautiful. Poetry can be inspiring. Poetry can be uplifting.

Poetry can even uncover your inner soul, which is God's gift to you.

5.0 Poetry for Meditations

Poetry as written by Omar is to be read carefully. It is suggested that the reader concentrate on the deep meaning of the verses, as related to survival, not only in a physical sense, but in a spiritual one as well.

Each of Omar's poems should then become a thing of spiritual beauty, with mental satisfaction, so that it draws a full visual picture in the reader's mind, according to his own experiences in life. Poetry can enlighten your life.

5.1 A Father's Prayer

Almighty God, I humbly offer my deepest soul felt gratitude,
For giving unto me the most treasured gift that I can receive or give,
That gift of being a father, helping to create another human being,
Brought to completion by my wife's loving care, with You Lord, providing the soul.

Almighty God, no higher privilege, gift, or wonder, can be sought or bestowed,
For life is empty without a child to provide for, meaningless, a vast void,
So, as I continue humanity's chain, formed by those who came before,
Remembering my child continues such a chain, aided by my loving care and guidance.

Never let me forget my duty to each child given to me, Lord,
As such shall last as long as my life on this earth,
Help me do my best for each, to be provider, protector, and adviser,
Nourishing each child's needs, aiding every step of their growth, during my life.

So Lord God, I ask Thee, indeed implore Thee, accept my deepest soul felt gratitude,
As I have striven always to be the father You wanted me to be,
Loving each child given to me, always grateful, always thankful, always remembering,
That deep in my soul, because of Thee, it hears the most wonderful sound, "Father …
".

5.2   Just Me

Into this world I came with nothing attached or brought with me,
Completely naked, unclothed, bare, crying and scared as you can be,
Only equipped with body, mind, and soul by parents and God was I,
Totally dependent upon all about me for years yet to come was me.

No title of royalty, entitlements none, with nothing to offer but me,
As years past my body, mind, and soul grew with each experience anew,
A childhood flew by as fast as allowed by all remember by just me,
An education in every aspect on ever subject deemed directed at ever level offered.

Some I liked, some I failed, but passed into a life of wonder, great joy, and suffering too,
And I was told that's life, keep going, duty, honor, country, responsibility, is for me,
So never I stopped, quit, or pause to long, just doing what must be done, moving forward,
Goals to met, tasks ot finished, always go on, never slipping or failure will win.

As life progressed I have not gathered titles, awards, fame, and fortune for me,
But I have helped all I could at the time I could, as duty, honor, country called for same,
And God's helping guiding hand has been present always when needed by me,
So when my call to leave this wonderful life comes, to join those who loved me so, I can gladly go.

———◆◇◆◇◆———

5.3 Perhaps Poetry

Perhaps, poetry is the language of God's angels,
Spoken in such a manner we cannot hear,
Always in a unheard but meaningful manner,
Full of caring, duty, responsible, deep meaning, and love for all.

Perhaps, our ears are not tuned to hear such poetic conversations,
Unless God Himself has a special message we need at a precise moment,
Then, and only then, at the moment chosen, are we allowed to do so,
And with great clarity are His messengers able to deliver same.

Perhaps, since we live in a mortal shell limited to a brief span, in God's eternal time,
It is best that God's angels deliver His messages in such a fashion,
For it is surely true that such beings are always busy with Heaven's work,
As there must be much to do, as we at times, seem not to hear His messages sent.

Perhaps, they are so beautiful in form, content, and just sheer unbelievable promise,
Since the Great Creator of our very eternal souls cares so much for each of us,
Or, since the angelic delivery of such utterances is so above our level of understanding,
We do not hear, since poetic deliverance's full meanings too often fall on deaf ears.

Perhaps, a fortunate few, here on this earth of ours, however at times unexpected,
Have, on rare occasion, been able to hear a bit of such poetic deliveries,
Thus, His message has been delivered, understood, and repeated for all to hear,
Placing upon everyone His message of total, complete, love for all under His care.

5.4 Behold and See

Don't you care at all,
All of you that pass by?
Look I say, just look again, I say,
Open your eyes, behold and see!

Don't you care at all,
All of you know of Jesus and His cross?
Look again, I say, just look again,
Open your heart, behold and see!

Don't you care at all,
As you walk your life's path?
Look once more, just once more,
Use your senses, behold and see!

Don't you care at all,
As your life flows past ?
Feel what your soul feels,
Your eternal self knows, behold and see!

Don't you care at all,
Since He gave you life, glorious life!
Now He gives you the wonders of death,
So you can live again, in His eternal care, behold and see!

5.5 With Pride, Take Care

Take care with use of your pride,
As it can turn and harm the wearer,
Quietly as a thief in the night,
Silently as the twilight comes.

Damage pride has done may not be noticed,
And cannot be undone at all,
Since it might take years to appear, if ever,
Or remain uncovered, time too gone, for any repair.

Harm by pride cannot be rectified,
Hurt to you and others cannot be removed,
Once done, it's done forever,
Never to be forgotten by giver, nor be forgiven by the receiver.

Hubris, nay pride, must be guarded against,
As it is like an echo in a mountain pass,
Reflected again, and again, and again, and again,
Take care with your use of pride.

5.6 Her Gentle Love, Given to Me

Love should be gentle, with a light touch,
Embodied and carried by a soul to match,
Almost as a gentle summer rain falls so very lightly,
Giving life itself to our garden of life.

Love must be given very freely, never ever demanded,
Expressed as a live spiritual thing of unmeasured value,
Almost as gentle rays of sunlight from providing our start,
Shining so very brightly upon our garden of life.

Love is an expression of our great God's gift of life itself,
Carried between each of us for the nourishment of our eternal souls,
Binding together in all measured ways our simple minds express,
Eternal thanks for all we have experienced in our garden of life.

Love is a bit of heaven we were all given at birth,
Sealed into each of us for later giving when needed,
Almost always freely seen at times of life's great needs,
Shinning forth, showing its beauty, fulfilling its promise, in our garden of life.

Love is what you have given me,
Watering my life and lighting my way,
Showering all of love's beauty into my life,
My mother, who gave me life, and her love always, always, . . .

5.7 With a glance Back

On this day of days but a glance back is made,
Remembering all of your beauty, radiance, and sheer love,
So freely given without a thought,
Remembered always with but a glance back, always, never forgotten

On this day of days but a glance is made,
Seeing all of your eternal beauty, spirit, and soul,
To be seen by all who have eyes and mind to see you so,
You will be remembered on this day, always, never forgotten.

On this day of days but a glance forward is made,
Seeing God's gift for you of life forever, with joy, and boundless love,
To be seen by all who love you so,
And with but a glance back, seeing you again, always, always, always …

5.8   Never Forgetting, Always Remembering

As the years melt into the vast abyss of endless time,
I remember you, young and beautiful, vibrant, radiant,
A beauty in body and soul, spirit and mind, a wonder to all,
And I remember, never forgetting, always remembering.

When each child came, life anew, celebration, happiness unbounded,
I remember you, as sunshine shone, love radiant, the miracle renewed,
More than a beauty in body and soul, but a woman in all her glory,
And I remember, never forgetting, always remembering.

As they grew, passing all stages from infant to youth to full mature human being,
I remember you, always helping, loving, listening to all, never leaving,
A saint, full of sacrifice, a protector, full of fury, but always full of love,
And I remember, never forgetting, always remembering.

Now the day seems to be done, the quest almost fulfilled, as each as gone his way,
I remember you, still seeking to help others, making a difference, contributing to all,
A mature beauty radiating grace, charm, wit, and boundless love, a wonder to all,
And I remember, never forgetting, always remembering,

Now that all seems to be done, as the vast abyss of endless time moves on,
I remember you, as my mate, completing my soul, giving my existence meaning,
Wise counselor, stout pillar, a unending wall of support through it all,
And I remember, never forgetting, always remembering, remembering, remembering, . . .

———◆◇◆◇◆———

5.9   A Whisper to Remember

Last night I heard a whisper from God Himself,
Laura is just fine here, with her family, in Heaven's embrace,
She walks again, no aid required, strong and straight,
Never again will she fall, hurt, or feel infected earthly pains, never again.

My guardian angels assigned shall always be near her,
Watchful, protecting, advising, ensuring her of your continuing love,
No harm will ever be allowed her in entire eternity to cross her path,
Laura has paid the price not even owed for such eternal bliss.

All her fears, concerns, and hurts have been banished, by My direct order,
No more examinations, medications, treatments, hospitals, surgeries,
Laura's heavenly body has been made from her pure spirit as perfect as can be,
Her enduring soul, courage, and determination are not lost but always here.

She is here, she will wait for you, when the Angel of Death will be sent again,
Bringing you back to her, God's whisper said so gently and quietly in my mind,
Telling me of Laura's continued love for me, her children, and her family,
All still on God's wonderful earth.

As the whisper from God Himself softly faded into the night,
I heard a familiar voice in the background say to all,
Live well, live strong, fear not, and embrace life as God's precious gift,
Remembering my love for you, and for all, and until we embrace once more,
I, am your Laura.

Note:
Composed by Laura Jo Holley Jackson's grateful husband of 53 years, 17 days, and
1.5 hours for having such a wonderful lady, wife, mate, friend, and soul making
living worthwhile.

5.10 An Instrument of God's Divine Plan

Lord, make me an instrument of Your divine vision for mankind,
Give me the direction, wit, and drive needed to help write Your divine plan,
Such that all now and after me all shall understand and strive to reach same,
Always reaching toward Your divine goals set for our humble race of souls.

Lord, forgive us all for our not understanding Your mighty insights given in holy texts,
All foreseen by Your gifted prophets sent so many ages ago for our enlightenment,
Misunderstood, denied, even crucified, due to our cruel inner natures exposed,
Including Christ Himself, our Savior, and Your only Son.

Lord, give me drive, energy, health and tools needed to help achieve Your noble vision,
Do not allow me to fail, stumble, falter, waste, or enter into obscurity before I contribute,
I pray to Thee always for direction to be Your instrument of delivery,
Allow me to aid, forgive me when I digress, and keep me on the right path, I pray.

5.11 My Guardian Angels

God assign's each upon birth one or on rare occasions two Guardian Angles,
Whose duties are to guard the soul, spirit, and eternal mind of the new one,
And even the physical body, yet to remain unseen by all during a lifetime,
But I know that I have had the extra need, and God did send two for me.

They have always keep close watch over and around my life,
As drastic and even fatal harm has always past me by to this very day,
Carrying shields issued by Heavens own armory's chief assigned,
And blessed swords that can fracture any evil sent to harm me.

Their names will never be known to no one on this earth,
Nor mostly likely never to even my soul, spirit, and eternal mind,
Until that moment I am called to join my beloved wife and all above,
To live the eternal life promised with all who loved and cherished me so.

Then as I give my accounting at the very gates of God's eternal domain,
And my "Book of Life" is read aloud and checked, and entry is thus assured,
As my Heavenly crown so hard earned from a life well spent is given to me,
I shall turn to my Guardian Angles and say "My eternal thanks and job well done."

5.12   A US Marine's Prayer

Listen to my prayer, Lord God, I beseech Thee, from this humble servant of Yours,
I must use weapons of war, upon those who bring nothing but fear, destruction and death,
Make me yours, my strength yours, so I can assure safety for all I am sworn to protect,
That fear, harm, hurt, and death I shall turn away from them.

Give me the unswerving will to inflict such destruction so fast and furious upon my foe,
That such foe shall fall before me, as if hit by all of nature's destructive forces at once,
So future foes shall quake at thought of facing Marines again in future times and places,
Never allowing harm to the innocents I protect and sworn to guard.

Lord God to You I have sworn to be faithful, the Marine Corps, my country, and family,
So when I must fight for right, always be at my side day and night, as battle rages about,
Keep me brave, give me needed warrior skills, stay at my side, no mater terrible the fight,
And if I die, lift my soul with Your heavenly embrace, so I may have Your eternal grace.

5.13 A Warriors Lament

Dear God I pray that You stand always on my side, not at my foe's,
But if not, I beg in my most humble way, move away from my foe's side,
Then and only then, stand by to witness the damnedest fight ever seen,
As I shall do all in my power to totally obliterate such foe from this earth.

And upon my victory hard won, I shall show all mercies following Your rules,
To the vanquished foe correct treatment and justice shall be served,
Knowing that then and only then, Your approval shall be mine,
And once more peace, justice, and security will be found, again for all.

———◆◆◆◆●———

5.14 Hidden Memories

Very deep in my soul I keep hidden memories of you,
Only I and I alone now know how wonderful life was with you,
With careful care and treading with the lightest step I dare enter there,
For such hidden memories stir my deepest emotions of you.

The dearest one in my entire life was, has been, and always will be you,
So such soul stored hidden memories must be treated with a tender touch,
Carefully unwrapped from its life remembered time and place of great joy,
Happiness remembered of you my dearest love to be savored beyond measure.

To quick as my time passes on I must move such a hidden memory back to it safe place,
So it is again safely put away deep into my hidden soul's storage place,
Never to be shared with others but only between God Himself and you,
As now you are in our savior's promised heaven forever under God's eternal watch.

I know your soul, spirit, and meaningful self are totally safe, secure under heaven's watch,
As my hidden memories of glorious moments of love, affection, and gratitude with
you,
Until we are united again within God's glorious grace, together for eternity to never part,
Such wonderful shinning hidden memories keep me living life's fullest embrace.

5.15 A Morning Greeting

May the radiance of a glorious new day shine brightly upon you,
Filling ever moment of the bright morning with happy thoughts,
With all of God's creations at peace from night's refreshing rest,
Your body, mind, emotions, and very soul looking forward to your day.

May your very soul be filled with blessings overflowing,
Spilling life's joys upon all spreading such happiness about,
May God's happiness be yours throughout the entire new day,
And may every morning, greeting every day, be the same in every way.

5.16 Bless Our Nation, O' God

O' God, we trust in Your divine mercy,
For only You can lift all misery from our land,
Thou are mighty indeed, above all men, rulers and the ruled,
Higher than the very sky itself, so we seek divine mercy.

Divine One, bless your meek and poor in this nation of ours,
Forgive all of us sinners, and lift your anger from us,
For we are but mortals, mere men of weak flesh,
Wrong as our very breath may be, have mercy upon us.

Heal the sick, mend the lame, protect the weak,
O' God, keep famine, war, and even death itself at bay,
Spread they kindness, shine they grace,
Forgive us all our sins, hear our lamentations, have mercy upon us.

For thou are the Mighty One, to whom we lift our voices,
We are singing praises, singing praises, always singing praises to you,
Our life, soul, indeed eternal existence itself is yours,
Its Your presence we humbly ask, so humbly ask, bless our nation, O' Lord.

5.17 Time Moves

As time moves, once again, and a loved one leaves again, and then again,
O' God, Most Magnificent One, cannot You delay time?
Cannot time be changed by its own Creator?
O' God, please, I beg Thee, tell us how to do so.

As time moves, again, and one departs us, once again,
O' Great One, Master of All, why is time, what is time, where is time,
Cannot time be recalled, brought back, made to obey, changed by willing it so?
0' God, why not divide time into two parallels or even many for us mortals?

As time moves, again, and another loved one from this life vanishes, I bow once more,
Before Thee, O' God of All, Supreme One, the only God I acknowledge,
Only bend, not break, time for me, and my beloved one,
For what does time mean to You, the Timeless One?"

As time moves, always again, and separates loved ones again, and again,
Only I cannot break, bend, nor ever forget times, sweet memories, gone by,
As only time itself, and the Supreme Creator, remembers why must it be so,
But time here will surely end, bringing loved ones together again in a timeless embrace.

5.18 For Our Sailor

May God's great hand be ever present to protect you,
May His grace shine upon you,
Keeping your sails full, on course, in the best of seas.

May God remember to watch over your small boat always,
As his seas are so very large, rough, danger always about,
Bringing you past all harm, safe always upon His oceans great.

May God always send a gentle rain, when it rains, with soft blowing breezes,
May God always let sunlight fall softly upon your ship,
And may God's stars guide you to a safe port, always in gentle seas.

5.19 A First Kiss

I hesitate with quaking mind and faint heart,
Dare ask a small favor be bestowed,
Upon a humble person's soul,
Asking from one with such grace, charm, and radiant beauty.

From such a one even a Greek God would not dare,
But I shall beg, plead, implore for such brief kiss be bestowed,
For such emotion, such transformation, such wonder, such feeling,
Will be transmitted by same to depths of my body, mind, and soul.

Leave me not knowing of you,
As a first kiss is often remembered in later years,
Not lost in the passing of so many long years,
Always remembered, never forgotten, by a humble person's soul.

5.20 My Wife

A song of love is always heard in your voice,
As my ear strains for each sparkling uttered sound,
To think I was your choice from so many suitors humbles me,
Nay, to hope, even pray that it always be so.

O' God, Most Merciful, thank You for giving me my ears,
So I may hear her voice, feel her spirit, listen to her very soul,
For she means all the heavens to me, just as stars give light,
Her sparkling voice is music to my being's core, need I say more.

5.21 Missing You

How much I miss you cannot be determine,
For it is as the same to count all of God's stars in His heavens.,
Or numbering the grains of sand on all the world's beach's,
And remembering all the wishes of all the lonely ever made.

Missing your very presence cuts into the depths of my soul,
Pouring its contents upon the sands of time, each grain lost never to be recovered,
Indeed my soul cannot be fulfilled, until we are together again my only love,
Under our Creator's eternal watchful care, never again apart from our completed self.

5.22 Spring Breezes

Spring breezes float softly by, like butterflies on their wings,
In the still of the evening as they so very welcome feel near,
And bring a scent of love to provide an aroma of life anew,
As senses once thought dead spring to life once more.

As a golden full moon floats across the eternal sky,
And as multitudes of stars shine down a radiance begun from long ago,
Even God's close planets stop spinning for a moment of time,
Because of spring breezes, reminders of long past, wonderful memories not forgot.

———— ◆+◆+◆+◆ ————

5.23 Never Alone

My eyes raised to the heavens,
And I felt fear,
For God himself was near,
And I was afraid, very afraid.

O' God, why am I alone I asked,
Must I always be so,
What have I done to deserve this nothingness,
Most High and Glorious Creator of all?

An answer came, as darkness grew,
With my eyes raised to the heavens,
As I felt so very afraid and alone,
Again afraid, afraid, and very alone.

Then I heard His night bird sing,
His song praising God all around,
With every single glorious note,
And then I knew what that meant to me.

My soul is to be filled again as alone I am not,
Never alone with such song in the air,
My memory stirs, and blessings occur,
For I have memories of you, the one God Himself sent to me.

O' Quite now always my fears be,
Nothing will nor can harm me again,
God's wonderful gift of life on this earth I have lived,
For I remembered, as now always, I can remember you.

5.24 Beauty

Not only is beauty "In the Eye of the Beholder",
But its recognition is hidden in the soul of the beholder,
Thus, it enables the eye to see beauty of many different types,
Life itself framed in a mother's love, a father's devotion, and nature's wonders.

Beauty is found in a glorious mountain pass, through highest peaks,
Towering above in glorious attempt to reach heaven itself,
Snow covered magnificent they are, reaching to the highest sky,
Seeming to just, but yet not quite, touch majestically infinity above.

Beauty is found upon a ocean gaze toward a far distant horizon,
Showing the earth's curve with its wondrous round gravity and mystery,
Putting sailors of the ocean deep under its quite spell as they see,
Rocking so gently their ship on the endless sea of life itself.

Beauty appears in a mother's concern and care of her children dear,
Always there, never lost, or spent carelessly, but available as needed during her life,
Given freely, never a charge or price upon same, never lost or given in shame,
Always to help, aid, comforting, soothing, reassuring with love abundant.

Beauty occurs around the clock of time, unexpected, but always welcomed just the same,
It is in a grandchild's smile, found in laughter, abundant in hugs, treasured by all,
Either giving or receiving it may be, always felt in a most wonderful way,
Adding flavor to life, improving its taste, smell, and feel to all who loving care.

Beauty is present in a clear night's sky, where our Creator hung countless stars,
All gloriously shining, placed for ever, writing a message for all to see,
The very cosmos itself in a never to be matched breathless eternal display,
Of God's mirrored never-ending reflection for all living souls to see.

5.25 Duty

Duty is doing what one is expected to do,
No matter what the circumstances are found to be,
Regardless of cost or consequences may be,
Yet the expected is accomplished as a must it be.

Duty is a sacrifice done without thought,
Since it expected as a must and cannot be bought,
Never delayed, put off, nor skipped but always done,
A task, one or more, accomplished without thought.

Duty is finished whenever the doer has carried same out,
No fanfare, no reward, nothing paid, just done about,
Help may or may not be contributed by others as well,
However its done, it never can be bought nor sold at all.

So duty is a must, always a must, finished at last,
By one or many, with no reward asked, nor ever given,
Because it comes in life as a necessary, at times unexpected,
Once duty is done, recognized, completed, it is done, life moves on.

———◆·◆·◆———

5.26 Honor

Honor is a quality each individual has deep within,
Given with his sacred oath upon his swearing same,
And all rely it being upheld, solid, never to be broken,
If else he does, his words will be seen as nonsense spoken.

Honor is a military member's guide and duty to uphold,
Since his brothers in arms, country, family, and freedom can be at stake,
Always it must be given and taken on one's very life,
Because all depend upon it, keeping life safe, as sure as heaven is above.

Honor given, above all else, is as solid a promise can be,
Never to be broken, as time passes by,
Similar to the Rock of Gibraltar, which cannot be moved,
Solid, sound, a granite foundation, never fractured, nor even cracked.

Honor between husband and wife, children and parents, family and all,
It is worth more than gold, precious gems, or treasures vast,
Never bought, traded, found, buried, mined, nor discovered,
However, always within reach of each, deep within, when needed.

Honor is given to all for this life's use,
To be spent carefully when occasion requires,
Never wasted nor thrown about as many may do,
For it must always be spent with care, thought, and never forgotten at all.

<hr>

5.27 Country

My country is founded upon great guarantees bought in blood,
Sacrifice, courage, determination, and a willingness never to quit,
Bold actions carried out in the face of odds sought insurmountable,
By men and women who never gave up, determined to win, and win they did.

My country was founded by patriots, who gave me freedoms of life,
Liberty, equality, worship choice of God Himself, and pursuit of happiness,
Life without fear, choices I make myself, levels of government all peopled by choice,
Not me for government, serving at a government's pleasure, instead, it serves me.

My country serves my interests, guaranteed by a Bill of Rights, a written Constitution,
And a representative legislature which I can freely choose, call upon, directed by all,
Once elected they can be removed, held accountable, dismissed from power too,
This is my country, were justice prevails, balanced for all, with juries required.

My country is a beacon for the entire world, strong, willing to fight for what's right,
Always full of people loving all of life's freedoms written down as a guarantee,
Protected for all by courts of law enforced, upon which life, family, and welfare depend,
Ensured always by a free, responsive courage of a determined people to stay always so.

————◆◈◆————

5.28 Questions on Life

Great Creator and God of All, please answer simple questions I humbly ask Thee,
As they follow now, my Lord and Savior, in such brief and short form of mine,
Some good people, wonderful they are, have to suffer so greatly in their life, why,
Some bad people, terrible in all ways, are allowed to harm others at will, why?

Many children, innocents, are deprived of basic necessities sorely needed, why,
Murder, robbery, theft, rape, torture, and other such depraved acts never stop, why,
Great trust placed in people in very high powerful positions oft go astray, why,
Many politicians seem to lie, cheat, steel, from society so easily done, why?

As each person travels through their life's journey down its unknown path,
Very few can every see such obstacles ahead, dangers, harms of any type,
Such must be avoided, gone around, solved with great effort, or great hurt occurs,
Such tests should be remembered, and taught to loved ones, is that there purpose?

Is life meant to beat some, those who cannot cope with pain inflicted at random times,
Or, if a person suffers, and then copes with such, is the lessons learned passed on?
Does strength of spirit then occur, with a more mature eternal soul as a result,
Building a better person in every way, for our eventual heavenly acceptance by You?

These questions and many more, cannot be answered by simple men alone,
Much greater minds, famous philosophers, theologians, and others have tried,
Always they fail, never can they answer such simple requests, simple questions,
So Great Creator and God of All, why is our life burdened so?

5.29 Some Answers, Maybe

Such answers to the simple questions posed must remain unanswered for now,
But when the Angel of Death gently touches my shoulder for my soul,
To carry me to the Gates of Heaven to enter same as Saint Peter allows,
My Book of Life shall be opened to all and weighted on God's scales by Him,
And answers will then be given in full measure to all my questions on my life lived.

Good people must suffer to experience difficulties less fortunate know in life,
So they know what needs to be done for others, to help provide better life for all,
Bad people will suffer as well, when the river to Hell's gate is crossed by them,
For they shall have an eternity to suffer, repent, wail, and cry for all misdeeds done,
In Heaven great joy is always known, but never in Hell, as the bad shall see.

Children, innocents, the old, lame, sick, and all who have suffered shall know,
In Heaven all will be provided, never to lack, love abundant, for all to have,
Illness none, souls made perfect after life's travails, all loved ones assured together again,
Eternity in God's care, as promised by Christ Jesus Himself, for all to share,
All criminal acts shall never be, with trust abundant, safety assured and known by all.

Thank God for life's problems, because each soul grows, as God means it so.

———◆▸◖◂◆———

5.30 When

When life's burdens seem ready to break one's very being apart,
A knee bent in heartfelt prayer uttered from deep in your soul must come,
Quietly sent to our Great Creator and God of All,
Carried for you by one of His angels on wings of heavenly gold.

Placed in His immediate in-box and already read in advance,
A return reply immediately delivered at lightening speed to you,
Fear not, despair not, for your thoughts and prayer have been heard,
Your life is not to be crushed, stopped, nor thrown away.

Indeed, no situation is ever without solution, no prayer overlooked,
Every act has a greater purpose, humanity's good shall always prevail,
Just do your part, try your best, always move forward, never quit, nor give up,
Remember I, your Creator and God, am always with you, your soul is part of mine.

Remembering this, your burdens in life are Mine as well,
Always they have been assigned with care and purpose,
And I never can be broken, by any force, energy, or combinations ever,
For I, your Creator and God, created all with purpose for all time.

5.31 Life's Basics

Only love forms a firm foundation for a life well lived,
Back by action, work, judgment, and firm accomplishment indeed when needed,
For love of life, for family, for right, for effort made, shall always count,
As basics of life are created for all to share as necessities are known.

Innovation, creation, action on thoughts, insights, inspiration as found,
Always thinking, moving, forward, with great victories, or small they may be,
In every effort, nothing should be wasted, time least of all, it alone cannot be replaced,
All forming a love of life, and a love for all, with freedom to do, under God's grace.

5.32 Bluebonnets

A field of bluebonnets in a early Texas spring radiates breathtaking beauty about,
It soothes the soul, brightens the spirit, reassuring one part of heaven is still around,
Another reminder of creation, showing for all to see, saying again a promise bold,
Blue as the sky azure above, fulfilling again nature's vast promises, Creator made.

Renewal once more, another beginning, life's march of time, onward toward infinity,
Onward it goes, here again as promised each spring, always in a beautiful way,
Each pedal perfect, together such forming breathtaking beauty, once more for all to see,
Heaven's deep blue, bluer than the sky can ever be, found each spring in vast Texas be.

Another start at life itself, as Texas bluebonnets proclaim that heaven is there for all,
Each field, with heavenly blue, shinning God's glorious promise once again,
Your forefathers saw such beauty, understood, and knew such was Creator sent,
Reminding all of what His heaven does contain, so Texas has a bit of heaven too.

5.33 Should, Could, Would

As I grow older weighted down by many years,
A few simple words ring over and over in my mind,
Over and over they sound, like a tolling bell,
"Should, Could, Would" are the few simple words.

Should rings loud and clear,
Questions they bring about actions not done,
Perhaps started, but finished poorly, if at all,
Should I have ..., should I have ... Should I have ...

Could is another one of the many heard rings from years gone by,
Over and over such a simple word sounds,
What could have been changed, altered, accomplished,
Could I have ..., could I have ..., could I have ...

Would rings through the years, perhaps loudest of all,
If I only would have done, would have done,
What changes in life's path would have been,
Would I been ..., would I saved ..., would I have ...

I shall never know life's answers to such simple words,
Now time has passed, long passed me by,
Never to be lived again in any way,
So "Should, Could, Would" are now left to others, as I pray.

5.34 A Morning Song

This morning as I started my day,
A beautiful song by a unknown bird began,
True notes high and sweet sounded clear,
As it's spirit soared with melody so dear.

As the sun itself rose in the early morning sky,
The songbird's tune rose in its beauty too,
O' hearing same was a blissful wonder to me,
As the bird's beautiful song began for me another day.

I asked myself about being alive on such a glorious day,
Why are such simple things, as a songbird's morning song heard,
Bring more joy to life than treasures of gold or gems of old,
Why is love given by a spouse, more valuable than life itself.

No answers came to such questions asked,
As the beautiful song was ended by the songbird itself,
The sun was up, the day had begun, and the enchanted moment over,
Alias, such simple moments are lost in time, but never forgotten, never ...

———————◆◆◆◆◆———————

5.35 Twilight

Into each life lived on this earth,
Under God's angels watchful eyes,
Toward our end years, called by many names,
But known by all, as twilight years, I come.

Slower in walk now for sure,
No longer running at all,
More pain in joints than ever before,
Sleep interrupted throughout the night, I am.

Carrying more weight than ever before,
Sitting in chairs more and more,
By choice day napping, because need says so,
My energies not charged, as before, I know.

Twilight of my life beckons now to me,
As I come under its careful shaded tree,
I am worn in body, mind, spirit, much older too,
Now my twilight is here, with angels near, I feel.

5.36 Rain

Falling from the clouds above, rain cleansing the good earth again,
It is really God's tears for us shed, renewing life's needs,
Gently it falls upon us all, giving such wonderful promise again,
As such gentle rains must fall, cleansing, purifying, and washing us all.

Sometimes rain comes in great gusts, blown by winds among us,
Brought by great storms, fiercer as fierce can be, darker than the darkest night,
Frightening as they are, they pass through as a danger in the night,
But God's displeasure crying such tears lasts, only brief moments in his sight.

However it comes, rains are sought for life sustaining needs, always, always, always,
So rain falling upon your life's path be gentle I pray, softly, nourishing you,
Its message meant to be God's most deep love, slowly, gently, falling about,
Especially for you, as His approval is so grandly shown, by His giving such gentle tears.

5.37 Mothers Are

Mothers give life, to each soul God places upon earth,
Mothers provide love, beginning for each child before birth,
Mothers give reassurance, through out each child's life,
Mothers advice is sought, as life encounters occur.

Mothers provide life, love, reassurance, advice, and always much more,
Mothers are supporting, encouraging, forgiving, with unmeasured love,
Mothers were made just as angels, so many mothers are angel like too,
Mothers are meant to be a fountain of love, comfort, and always strength for all.

So we thank our God, for our mothers, His enduring gift to us all.

———●◆※◆●———

5.38 The Unexpected Storm

When unexpected rain falls upon your path of life,
Always gentle may it be,
Washing away all your troubles carefully.

And when sunshine once again makes your path dry,
Know again all is made right as can be,
And may God's angels continue watching over you.

5.39 A Day Begins

As I watch the early rays of heaven's sun emerge,
Quietly, so very, very quietly, not even the babe stirs,
Nor flower petals of the rose move,
No response, no response, so very, very quietly it comes.

But all begins to gently change, as another, then another ray joins,
For life itself must stir, as the great orb, blinding as it is soon to become,
Peaks above the mountains so high,
Waking all, but not with cries, just quietly, so very quietly.

Soon all of the earth is aware, another day has arrived,
To be used, joyfully, carefully, wondrously, lived by all,
Another gift from the Most High, another day, another day,
For all to do with what they will, a treasured gift, so priceless, so rare.

For all life now is astir, never bothering to count how many rays it takes,
Nor how much silence each breaks,
For such a treasured gift bestowed upon all life,
By our Creator who walks, so quietly, so very quietly, just like His rising sun.

———◆+◆◆◆+◆———

5.40 Life's Seasons

Life starts long before birth, as a very spiritual beginning,
From God's 'Well of Souls' directing your soul to it's earthly journey,
Melding into a preformed wondrous awaiting living vessel, your body,
Formed as a small helpless human, provided with emotion and mind.

The total you bursts down the birth channel, a major miracle seen again,
It is the very beginning of your life's first season, it's God's gift,
Your Spring begins, O' so full of life, loved, growing, playing, everything so glorious,
Unfolding mysteries, crawling, walking, talking, and dreaming unknowns.

No worries, responsibilities, cares, or troubles whatsoever are yours at all,
Just sleep, growth, progress each day with much love, care, and great joy for all,
Learning at every moment, developing all senses to the fullest, and now thinking starts,
Walking, blessed speech, and knowledge stored, so very quickly for all.

Soon childhood is achieved, including play with others, as social skills are learned,
Early friends are found, but not remembered at all, as time rushes you onward,
Physical skills develop, balance perfected, and emotions emerge from you,
A personality develops, from a deep combination of soul, mind, and body; its You.

Alias, Spring is over as early Summer arrives, with a jolt, and formal schooling with same,
Its kindergarten, first grade, quickly followed by others, conferring basic knowledge,
Summer is a vast learning of facts, forcing your mental abilities at times to the max,
But its a kind season, with growth beyond measure, in ever human dimension known.

Mid-Summer begins as you enter years know as "teens",
O' joyful times abound, as teen years bounce around,
In which a condition called "maturity" is sought, but rarely found, nor sought, or taught,
So the quest now starts for body, mind, and soul as expansion into a young adult begins,
But temptations to fail are many, and we all do, in one manner or another; yet its You.

Omar

Mid-Summer teaches one to overcome any failures, great or small, as character is built,
One's strength comes from struggles overcome, whose great teacher is called "failure",
Your growth continues as Late-Summer begins in what is now called "High School",
Where you are told "Choose a lifetime's trade, profession, skill now, now", so
You must.

Late-Summer continues as last teen year occurs, with early twenty years to soon found,
College, military service, work starts, within Late-Summer's shinning sphere,
And, with great luck, just perhaps, God introduces a most wonderful mate of the other sex,
Most gloriously enchanting, perfect in all wonderful aspects, carrying a lifetime's love; for you.

Summer over, very early Fall arrives, marriage, children perhaps, responsibilities galore,
Real lessons in life begin earnest now, work, multiple needs of family, obligations to met,
Never enough time, funds, tools to cover all, life's full enjoyment beyond measure yours,
Love abundant, growth measured, success of all within reach, challenges met; life for you.

Mid-Fall arrives, life's cup overflows, family, love for all, with warm glorious emotions,
Accomplishments abound, challenges are fully met, the taste of life full measured for all,
Maturity occurs within, and without, and about, goals are complete, happiness is yours,
Late-Fall is entered, your very soul shouts out, life you have known completing; for you.

Early-Winter, as "Your Golden Days" start, announced by slower step, an added caution,
Perhaps sensed ache never felt before, a milestone age, called retirement; just for you.
Mid-Winter arrives soon after, not really wanted, comes like a thief at night, not invited,
But it cannot be stopped, as time progresses, never turns back, each season must be;

Late-Winter here, "Your Golden Days", memories shared, family, love, maturity found,
With infinity yet to start for your soul's return to God, whenever God requires it back,
Challenges overcome, assigned work accomplished, your life to leave a better world,
Because your efforts, love, sacrifices made, helping all, caring always; its you.

Late-winter now allows joy of life's wonders, music, remembrances, love, of caring souls,
Whose life's paths crossed yours, poetry of living, language spoken, thoughts created,
Freedoms found, literature's read, eternal soul now clearly seen, of one's married mate,
Now Late-Winter time allows needed wisdom backward and forward on time's
final line.

Soon God will send His souls harvester, called many names, but fear not, for
God does so,
With grace, redemption, and Christ Jesus promise of old, your place assured in
His heavenly abode,
Once more united with all loved ones, a eternity of great joy, all cares banished,
Immeasurable love found, with your soul mate in eternity assured, God's promise
for you!

5.41 Dreams on Life's Path

Dreams once occurring are stored in your Dreamland,
Never to be lost or forgotten, but always recalled as needed,
Remembered, but not remembered, as life itself moves forward,
Not lost, not really meaningful, during the present, but always available.

Dreams start before your journey down the birth channel into life itself,
Keeping you safe, secured in your mother's love, so very cared for,
But suddenly your slumbers are only yours forever more,
As the soul is inserted by God himself during nature's birth process.

Now as a babe your dreams are yours, and yours alone for evermore,
As you follow your given genes abilities life's chance has imbedded in you,
Childhood dreams being once in a lifetime, such as cowboy, Indian, other hero beings,
Others always in your growing mind, changing into other unattainable dreams.

In older youth, dreams come from first loves, other first glorious experiences,
Some forgotten, some remembered, more mature dreams filtering life for you,
Until some form of maturity drive some of your dreams into goals to be achieved,
Failures occur, but successes too, as dreams help shape your life into reality itself.

In full maturity of years, some dreams fulfilled, many not, but dreams are but dreams,
However, life shaped by circumstance, chaos, unseen always, but dreams are not forgot,
So you wish upon a star so far, far, away, hoping a dream unfulfilled can still come true,
Fully knowing it is but a wish, a hope, a distant desire, never to be, never at all to be.

As life closes upon you, it must for all living things, knowing no exceptions ever allowed,
Dreams can be recalled, making life's lived more abundant in love, kind acts, good deeds,
All remembered, as perhaps dreams, but perhaps real as well, and can be taken with you,
Into God's heaven, through heaven's gate, with Saint Peter's blessing to share in eternity.

Heaven is not a dream, but a promise, from Jesus Christ, God Himself in another form,
That your eternal final dream is true as sunlight, earth, water, and air we live with today,
Your loved ones already there await with eternal joy, your appointed time to arrive,
So your final dream is not lost, misplaced, or failed, but an eternity with love ones is true.

5.42 A Daughter's Valentine

From the time I first saw you, my only daughter, my heart glowed,
Indeed, my heart beat with pride that God gave me such a perfect gift,
A special life provide by your mother, with God supplying your soul,
Perfection, beauty, life, provided again once more for my protective care.

Over the years your mother and I watched with pride your many challenges,
All conquered with grace, charm, and common sense, all in ample supply,
As you met life on all its varied fields of challenge, a winner every time,
No circumstances unexpected, unforeseen, nor setback, every defeated you.

Always forward on your life's path, always positive, with great assurance,
Forward you move, adding experiences for positive gain, growing, maturing,
Now a mature person, a great teacher of the young, molding their life's,
Shaping futures yet to unfold, building a better life for all, yet to be seen.

My only daughter, small as you were, now a giant in soul and spirit too,
Your mother and I, and surly God too, are so very pleased that our love,
Indeed, all family, all fiends, all those that have, and will love you know,
Are very pleased to ask you, to be our special Valentine, on this special day.

5.43 The Whip-poor-will's Nocturnal Song

During the early Texas Spring, just as the sun ceases its shining day,
   If you listen carefully, very carefully, during this twilight time of evening,
A most beautiful meaningful bird's song of crying for its companion may be heard,
   Very quietly, sounding its beautiful melody of its soul's longing needs.

Its beauty of sound, carrying its mournful plea of deep need, is quietly heard,
So beautiful sounded that all nature stands still, very still, listing to such a plea,
   Only heard, early in a Texas Spring's close of day, just a small echoed sound,
Calling 'whippoorwill', 'whippoorwill', carrying its lonely meaning to your heart,

Yet so quite in volume, so hard to hear, even your heart sounds must wait a bit,
So resounding quite, as it asks within its cry, "Where are you?", "Where are you?",
   And again it may repeat its mournful song, saying "I need you", "I need you",
Calling again, 'whippoorwill', 'whippoorwill', carrying its meaning over an over.

As you strain to hears its lone quite sound again, all of nature stops as well,
Life itself seems to know, including the early night stars now waiting to shine,
   Once more its mournful song is sounded, "My soul hurts", "My soul hurts"
Calling again, 'whippoorwill', 'whippoorwill', carrying its meaning clearly to you.

Finally, in a beautiful way, its last cry is barely sounded, as its evening song ends,
As you hear a soft reply, "Our love forever", "Our love lives", so very clear,
   Followed by a answering, "My soul knows.", "My love always.",
Then a final response, 'whippoorwill', 'whippoorwill', filling all its soul's needs.

5.44 My Plea as I Grow Old

Remember, I promised, if I lived to grow old, I still need you, forever,
Your spirit, charm, beauty, and love of life shall always be with me,
As my memory of you shall never fade, dim, nor be forgotten,
Not of you, my sweet one, as it is imbedded so very deep in my heart.

Always remembered, never to be forgotten, no matter what else occurs,
As the very world can cease to be, but not such memories of our years together,
Our life and years will always be remembered, as a love never to be lost,
Never of you, my dearest one, never lost or misplaced, not for even a brief moment.

Your smile, voice, touch, all such dearly remembered, all so freely given,
Your presence, always needed, felt, seen, your love for me, indeed for all,
A great lady's wondrous spirit personified, with such grace and charm,
All you, only you, given without thought, in such a wondrous way.

Now you have left me behind, called into God's heavenly kingdom first,
Granted your eternal rest, from all of your earthly life's trials and troubles,
To spend in God's eternity, with many of your loved ones already there,
Yet I know that you wait for me, as you have, and will, always remember me.

So remember me my love, as my time grows shorter by the day,
Yet I have more to do before I come to you, to share eternity together,
And since God has a bit more for me to do, stand by in eternity for now,
Just remember me as I grow old, soon again, I shall be with you, my love, my love.

5.45 What Words Can Do

Take care with words spoken, written, remembered, used without care,
Remember such can cause great pain, harm, consequences that one must bare,
Once spoken, uttered in haste, without thought, such may never be forgotten,
Only remembered, embedded in one's memories, not yet abandoned, nor lost.

Words can praise, express hope, love, success, achievements, and be glorious,
Conveying great ideals, tremendous knowledge, give life brilliant meaning,
Raising individuals, groups, even nations, to achieve wondrous achievements,
Expressing beautiful human emotions, love, kindness, mercy, grace, unbounded.

Indeed, words are more powerful than we know, carrying our thoughts to all,
Traversing time itself, when written down, clearly, concisely, so effortless,
As future unknown generations may read, understand, ignore, or accept same,
Clearly, words travel through time itself, into time beyond one who wrote them.

Knowledge, information, instructions, descriptions, hard won over generations,
Beauty, philosophy, religion, history, science, engineering, all are conveyed,
By words, so very important, expanding human thoughts to future generations,
What words do, or should not do, cannot be known in one's own time, so use
carefully.

5.46 Christmas in Moscow

T'was the night before Christmas,
And all through her apartment,
Nothing was stirring, not even a Lisa,
As she was fast asleep dreaming,
Of sugar plums, pudding, and Russian borsch.

As usual it was very below freezing cold outside,
Snow, ice, and all cold that was not nice,
Still her dreams were warm and cozy as a fireplace fire,
Keeping her warm as toast deep in her dreamland bliss.

Suddenly in the middle of a snowy Moscow winter night,
A loud knocking at her door was waking her up,
A loud voice demanded in for a hot vodka tonic,
Along with smaller voices saying the same.

As Lisa put her house robe about her chilled body,
Slipped on her Russian bear skin house shoes,
And stomped to the door demanding,
Who, why, and what was at her door.

An answer came loud and clear,
It is I, Father Christmas, and all his elf helpers,
Come to bring presents to a lost Texas great lady,
As well as some Texas cheer on this eve's night.

After all you deserve a great big "Howdy Y'All",
Especially brought directly to you,
From all family, friends, and others that love you so,
On this Christmas Day they sent a "Very Merry Christmas"!

5.47 To Live Forever Here?

I have thought what it would be to live forever on this earth, Lord God Almighty,
And after much thought, Dear God of All, I believe that Your divine limit is indeed best,
Why, is asked of me, and I have an answer to offer unto all, Dear God, as follows below,
Such answer might seem shallow, limited, weak, unreasoned, even faulty to others.

Time is limited, for any human given life on this earthly sphere, because You made it so,
However, with great reason, as continuing life's struggles forever here is not living,
But, instead, it would be a sentence to everlasting torment of body and soul on earth,
As such, a human would only experience terrible suffering, far beyond our capacities.

Each would see his loved ones live and die, while he continued onward on a timeless path,
As valued loved ones suffered the stings of life encountered, witnessed by a timeless one,
His sorrow, helplessness to intervene, as death itself claimed each valued loved one,
Such would bring un-surmountable pain to his body, mind, emotions, and eternal soul.

Life itself, forever on this earth, would then become an unbearable burden to endure,
So his eternal soul would pray to be allowed entry through Your eternal heavenly gates,
Where his loved ones of his life's time are most certainly already present, waiting for him,
Leaving the gift of living on earth forever far behind, as he has lived out his allotted time.

In his Book of Life, is to be found, doing even more that You requested of him,
His completed assignments You, Dear God of All, placed at his birth,
All done, well done, completed during his life's struggles with all such tasks as listed,
And, his most loved one will say, as he enters heaven, "Welcome, My Eternal Love".

6.0 Additional Poems for Meditations

Addition poems that follow in this section are composed by Omar's daughter, and help illustrate that poetic talent, as any gift, may be passed via the gene pools we all carry. However, this claim may be discussed, and even disputed, by many (or all for that matter).

But meaningful poetry it is, so please read and then mediate upon same. It might refresh your own emotions, and provide a uplift to know that following generations are loaded with all sorts of talents and abilities. Most probably never fully recognized.

Our human race has improved over the centuries since mankind's civilized history begun. Such improvement does offer all of us great hope for the future generations.

Great hope indeed.

6.1 A Daughter's Poem for Her Dad

I know I am not a democrat,
My vote is for to right.
My answers are frightfully pat,
My aspirations out of sight.

You don't understand why I can't cook,
And detest the way I drive.
You're uncertain of my red-headed look,
And wonder how I stay alive.

You've never met many of my friends,
And wonder how I spend my time.
Does my money stretch to meet my ends?
Are my eating habits really a crime?

When will I finally get hitched?
And do I really go to church?
Which big dreams have I ditched?
What men have I left in the lurch?

Who is this stranger who walks in your house,
And turns it upside down?
Who some times makes you feel like a louse,
And other times a clown?

I am your daughter, Father dear,
The one you bounced on your knee.
I am the one you held so near,
That re-headed stranger is me.

Gee Dad, it seems we've hardly met,
We are two strangers here.
Out of all the things that I could get in life,
You are one of the most dear.

So, thank you Dad, for all you have done,
And for all you've been for me.
I'll never be your favorite son,
But God help us, I'll always be me; Love, your daughter!

6.2 The Little Man I Saw in a Wood

I see him through the mists of night,
Busying himself before morning light.

He smokes a pipe and smiles a smile,
A smile that grows bigger all the while.

He has a funny walk and a body short an stout,
I like to watch him laugh and sing, and run round about.

The little creatures awaken, through the sleep wood,
And join the little leprechaun in his happy mood.

Soon the dancing and singing stops, for it is nearing dawn,
And we have to say goodbye to our little leprechaun.

He smiles at everyone, with his little smile,
And says he hopes to see us in just a little while,

Then he walks away into the shadowed wood,
And we stare regretfully, into the place where he had stood.

A leprechaun is a person few mortal men ever see,
But I often dream of the night when one was seen by me!

—————◆◆◆◆◆—————

6.3 Love Is

Love is here today and here tomorrow,
It may cause you happiness, or it may cause you sorrow.
It may bring you laughter, or it may bring you tears,
But it will stay with you through the years.

Love has no shape, or color, or hue,
Because it's an invisible part of you.
Some people say that love is a seed,
That grows with each loving thought or deed.

Love is something that you share,
To show your loved one that you care.
It is sweet and kind and good,
But often mistreated and misunderstood.

Love comes down from the heart and not the head,
It lives in most of us, but in some it is dead.
Those people are a sad, empty shell,
Instead of a heaven, they live in a hell.

Love is a mystery that baffles man,
They have tried to solve it since time began.
But it is a gift from heaven above,
Therefore, there is no explanation for this thing we call love!

6.4   Where is Darkness in Death?

Though darkness is at your doorstep and God seems far away,
Remember he is with you with every passing day.

When the pain becomes unbearable and you feel that you are about to cry,
Remember it is a beginning, not an ending to die.

When your loved one is in heaven, he'll be happier that before,
For he will be safe and happy in God's hands for evermore.

And when he is in heaven, even if you feel sad and blue,
Remember he is happy, and his spirit will always be with you!

———————●◆►◄●———————

7.0 Songs of Omar — His Poetry

The following poems were translated from ancient documents believed to be written by Omar the Insignificant, during a later lifetime of his. Much later, as this prayer was written at the time of Islam's beginning. So instead of God in his later poems you will see his use of Allah to address the Creator. Some experts still debate his authorship as translators and finds at different "digs" occur.

Note: What follows is thought to be poetry taken from the same documents. The reader is reminded that translation work is still underway. Refinements will be made later.

7.1   O' Allah, Remember Us Slaves

Least forget, O' Allah, ours is to work, not rest,
Ours is to struggle, to pull, push, and strain,
Ours is to serve, to grovel, to bend, and bow,
We are not to think, dream, hope, at best a wish upon a star is ours.

Least we for get, O' Allah, ours is to submit, not question,
The master's whip is quick, even deadly at times as the asp,
We must not forget, always, the master's ego, to feed and pamper it,
Never offend, keeping the truth hidden, the tongue silent,
At best, our salvation is as moonlight on the desert sands.

Least, we forget, O' Allah, ours is to endure, yet not cry out,
Ours is to dream, not achieve joy or paradise too soon, and not as slaves,
Ours is to beg, plead, seek mercy, not to demand nor take,
Nothing is ours, only the illusion exists, as our life ebbs away.

7.2   Ode to Slaves

At best pain will go, bad will vanish, and Paradise will be ours.
Slaves make the world move, while Master's make the slave toil,
Hang head low, advert the eyes, dare not speak the truth, make the slave toil,
Move, tote, work, ask not from the Maser, make the slave toil,
Waste, worry, agree, hurry, make the slave toil.

Do not, stand tall, bend, carry, be wary, make the slave toil,
Think not, reason be gone, Master's way or not, make the slave toil.
Slave. Toil, Slave toil, Slave toil, Slave toil; toiler is the slave,
The master reaps, the slave weeps.

Note: The previous poem illustrates survival under slavery, but nothing else.

7.3   A Day Begins

As I watch the early rays of heaven's sun emerge,
Quietly, so very, very quietly, not even the babe stirs,
Nor flower petals of the rose move,
No response, no response, so very, very quietly it comes.

But all begins to gently change, as another, then another ray joins,
For life itself must stir, as the great orb, blinding as it is soon to become,
Peaks above the mountains so high,
Waking all, but not with cries, just quietly, so very quietly.

Soon all of the earth is aware, another day has arrived,
To be used, joyfully, carefully, wondrously, lived by all,
Another gift from the Most High, another day, another day,
For all to do with what they will, a treasured gift, so priceless, so rare.

For all life now is astir, never bothering to count how many rays it takes,
Nor how much silence each breaks,
For such a treasured gift bestowed upon all life,
By our Creator who walks, so quietly, so very quietly, just as the sun rises.

7.4 Song of Omar – Life's Seasons

Note: Once again uncovered ancient scrolls have been translated by experts, to the best of their ability.

Once more our literature is to be vastly enriched by another poetic addition to the poetry collection called "Songs of Omar", believed composed over 3000 years ago by the ancient teacher, scribe, scholar, statesman, and philosopher, Omar the Subjugated, servant to
David the King.

In these new verses, Omar the Miserable One, seems to be praising life itself in the most glorious phrases since the Great Bard (Shakespeare, to you readers).

Omar's prose indeed can be very moving. We let the reader now judge for him/her/yourself (no gender selected nor implied these days).

The latest additions to "The Songs of Omar" now follow:

Song of Omar, Life's Seasons

Life starts long before birth, as a very spiritual beginning,
From God's 'Well of Souls' directing your soul to it's earthly journey,
Melding into a preformed wondrous awaiting living vessel, your body,
Formed as a small helpless human, provided with emotion and mind.

The total you bursts down the birth channel, a major miracle seen again,
It is the very beginning of your life's first season, it's God's gift,
Your Spring begins, O' so full of life, loved, growing, playing, everything so glorious,
Unfolding mysteries, crawling, walking, talking, and dreaming unknowns.

No worries, responsibilities, cares, or troubles whatsoever are yours at all,
Just sleep, growth, progress each day with much love, care, and great joy for all,
Learning at every moment, developing all senses to the fullest, and now thinking starts,
Walking, blessed speech, and knowledge stored, so very quickly for all.

Soon childhood is achieved, including play with others, as social skills are learned,
Early friends are found, but not remembered at all, as time rushes you onward,
Physical skills develop, balance perfected, and emotions emerge from you,
A personality develops, from a deep combination of soul, mind, and body; its You.

Omar

Alias, Spring is over as early Summer arrives, with a jolt, and formal schooling with same,

Its kindergarten, first grade, quickly followed by others, conferring basic knowledge,
Summer is a vast learning of facts, forcing your mental abilities at times to the max,
But its a kind season, with growth beyond measure, in ever human dimension known.

Mid-Summer begins as you enter years know as "teens",
O' joyful times abound, as teen years bounce around,
In which a condition called "maturity" is sought, but rarely found, nor sought, or taught,
So the quest now starts for body, mind, and soul as expansion into a young adult begins,
But temptations to fail are many, and we all do, in one manner or another; yet its You.

Mid-Summer teaches one to overcome any failures, great or small, as character is built,
One's strength comes from struggles overcome, whose great teacher is called "failure",
Your growth continues as Late-Summer begins in what is now called "High School",
Where you are told "Choose a lifetime's trade, profession, skill now, now", so You must.

Late-Summer continues as last teen year occurs, with early twenty years to soon found,
College, military service, work starts, within Late-Summer's shinning sphere,
And, with great luck, just perhaps, God introduces a most wonderful mate of the other sex,
Most gloriously enchanting, perfect in all wonderful aspects, carrying a lifetime's love; for you.

Summer over, very early Fall arrives, marriage, children perhaps, responsibilities galore,
Real lessons in life begin earnest now, work, multiple needs of family, obligations to met,
Never enough time, funds, tools to cover all, life's full enjoyment beyond measure yours,
Love abundant, growth measured, success of all within reach, challenges met; life for you.

Mid-Fall arrives, life's cup overflows, family, love for all, with warm glorious emotions,
Accomplishments abound, challenges are fully met, the taste of life full measured for all,
Maturity occurs within, and without, and about, goals are complete, happiness is yours,
Late-Fall is entered, your very soul shouts out, life you have known completing; for you.

Early-Winter, as "Your Golden Days" start, announced by slower step, an added caution,
Perhaps sensed ache never felt before, a milestone age, called retirement; just for you.
Mid-Winter arrives soon after, not really wanted, comes like a thief at night, not invited,
But it cannot be stopped, as time progresses, never turns back, each season must be;

Late-Winter here, "Your Golden Days", memories shared, family, love, maturity found,
With infinity yet to start for your soul's return to God, whenever God requires it back,
Challenges overcome, assigned work accomplished, your life to leave a better world,
Because your efforts, love, sacrifices made, helping all, caring always; its you.

Late-winter now allows joy of life's wonders, music, remembrances, love, of caring souls,
Whose life's paths crossed yours, poetry of living, language spoken, thoughts created,
Freedoms found, literature's read, eternal soul now clearly seen, of one's married mate,
Now Late-Winter time allows needed wisdom backward and forward on time's final line.

Soon God will send His souls harvester, called many names, but fear not, for God does so,
With grace, redemption, and Christ Jesus promise of old, your place assured in His heavenly abode,
Once more united with all loved ones, a eternity of great joy, all cares banished,
Immeasurable love found, with your soul mate in eternity assured, God's promise for you!

Note:
The previous poem contains the authors thoughts on the four major stages of a human beings life on SOL3, the third planet from the star called Sol, found on the far edge of a spiral of the Milky Way Galaxy, located only four light years from the Andromeda Galaxy which is our Milky Way's closest galactic neighbor.

The readers exact position in the known universe can be determined by the reader using this data. It is a straight forward mathematical exercise in celestial mechanics, and then GPS usage. The exact position of God Himself is more easily located, finding a bit of Him in each of us, with the rest located in His eternal being.

8.0 Understanding "Omar's Guide for Surviving this Turbulent Age"

This section is an attempt to help clarify terms and definitions as used in Omar's authored guide. It contains terms, definitions, noted people mentioned, and finally background letters written by Omar himself.

8.1 Terms

General Relativity: The laws of physics covering the large scale of the universe as described by Einstein's General Theory of Relativity (sometimes referred to as point particle theory).

Quantum Mechanics: The laws of physics covering the smallest scale of the universe's very basic building blocks that Einstein's General Theory of Relativity does not cover. This includes String Theory, and Multi-String Theory (sometimes referred to as string energy theory), as well as Black Holes. It occurs at the Planck length of the space-time frame. Fredrick Planck was a noted physicist who postulated it originally, followed later by Nicholas Bohr,

8.2 Definitions

CIA:

Central Intelligence Agency (CIA) is one of the principal intelligence-gathering agencies of the United States federal government.

ENIAC

ENIAC (Electronic Numerical Integrator And Computer) was the first electronic general-purpose computer. It was Turing-complete, digital, and capable of being reprogrammed to solve a large class of numerical problems.

FBI:

Federal Bureau of Investigation (FBI) is a governmental agency belonging to the United States Department of Justice that serves as both a federal criminal investigative body and an internal intelligence agency (counterintelligence).

Nom De Plum:

A pen name, nom de plume, or literary double, is a pseudonym adopted by an author. The author's real name may be known to only the publisher, or may come to be common knowledge.

SOL:

Astronomical name for our star (the Sun). It is the Latin name of the main-sequence star of the synonymous Sol System, called the plain Sun in English.

SOL3:

Astronomical name assigned to our planet, the third from the Sun, named Earth, in non-astronomical terms.

STEM:

STEM is an acronym referring to the academic disciplines of science, technology, engineering, and mathematics.

8.3 People Mentioned

Carter, Jimmy:

Former president of the United States. James Earl "Jimmy" Carter, Jr. (born October 1, 1924) is an American politician and member of the Democratic Party who served as the 39th President of the United States from 1977 to 1981.

Locke, John:

An English philosopher and physician. John Locke (29 August 1632 – 28 October 1704) is regarded as one of the most influential of Enlightenment thinkers and was the originator of the term "constructive destruction", which is one of the pillars of the capitalistic economy, that is the constant replacement of some need by a newer better improved approach, i.e., pony express, progressively replaced by the telegraph, telephone, short wave radio, cell phone, and now the Apple "wrist watch communicator. This "constructive destruction" applies to all fields of the capitalistic economy. His writings influenced Adam Smith, author of "The Wealth of Nations" in 1776 on the capitalist economy.

Christ, Jesus:

Founder of the Christian religion. Jesus (7–2 BC to 30–33 AD), also referred to as Jesus of Nazareth, is the central figure of Christianity, whom the teachings of most Christian denominations hold to be the Son of God. Christianity regards Jesus as the awaited Messiah of the Old Testament and refers to him as Jesus Christ, a name that is also used in non-Christian contexts.

Bohr, Niels:	Niels Henrik David Bohr (7 October 1885 – 18 November 1962) was a Danish physicist who made foundational contributions to understanding atomic structure and quantum theory, for which he received the Nobel Prize in Physics in 1922.Noted theoretical physicist and researcher in Quantum Mechanics.
David, King:	God's anointed King of Israel; the second king of the United Kingdom of Israel and Judah, and according to the New Testament Gospels of Matthew and Luke, an ancestor of Jesus. His life is conventionally dated to c. 1040–970 BC, his reign over Judah c. 1010–1002 BC, and his reign over the United Kingdom c. 1002–970 BC.
Einstein, Albert:	Albert Einstein (14 March 1879 – 18 April 1955) was a German-born theoretical physicist and philosopher of science. He developed the general theory of relativity, one of the two pillars of modern physics (alongside quantum mechanics) based upon space-time relational concepts.
Planck, Max:	Max Karl Ernst Ludwig Planck (April 23, 1858 – October 4, 1947) was a German theoretical physicist who originated quantum theory, which won him the Nobel Prize in Physics in 1918.
Bacon, Francis:	Sir Francis Bacon was a scientist, philosopher, courtier, diplomat, essayist, historian and successful politician, who served as Solicitor General (1607), Attorney General (1613) and Lord Chancellor (1618) in England.

Confucius:	Confucius (551–479 BC) was a Chinese teacher, editor, politician, and philosopher of the Spring and Autumn period of Chinese history.
Dickinson, Emily:	Emily Elizabeth Dickinson (December 10, 1830 – May 15, 1886) was an American poet.
Dumas, Alexander:	Alexandre Dumas, born Dumas Davy de la Pailleterie (24 July 1802 – 5 December 1870), also known as Alexandre Dumas, was a French writer.
Homer:	Homer (Ancient Greek around 850BC) is the traditionally-credited creator of the Iliad and the Odyssey, revered as the greatest of Greek epic poets.
Joyce, James:	James Augustine Aloysius Joyce (2 February 1882 – 13 January 1941) was an Irish novelist and poet.
Puskin, Alexander:	Alexander Sergeyevich Pushkin (6 June,1799 – 10 January, 1837) was a Russian author who is considered by many to be the greatest Russian poet and the founder of modern Russian literature.
Scott, Walter:	Sir Walter Scott, 1st Baronet, FRSE (15 August 1771 – 21 September 1832) was a Scottish historical novelist, playwright, and poet.
Smith, Adam:	Smith (16 June 1723 – 17 July 1790) is known as the "father of modern economics". His works are still among the most read in the field of economics (reference "The Wealth of Nations"). He is considered the father of modern economics.

Voltaire, Francois: François-Marie Arouet (21 November 1694 – 30 May 1778), known by his nom de plume Voltaire, was a French Enlightenment writer, historian and philosopher famous for his wit, his attacks on the established Catholic Church, and his advocacy of freedom of religion, freedom of expression, and separation of church and state.

8.4   Historical Background of Omar

Rumors, suppositions, hearsay, historian's opinions, etc … exists in abundance on the origin of our author, Omar 'The Most Magnificent'. Was he a scribe in King David's court? Or perhaps the Chief Musician who helped write the Book of Psalms? Did he help in formulating King Solomon's wise sayings? Has he been given more life's than the always allocated one?

Was Omar 'The Wise All Conquering General', a known warrior? Was Omar a great scholar? Or was Omar just a slave during ancient times? Is Omar still with us? How could a mere man live 3,000 years?

Is Omar a alien hidden among us? If so, from what location and for what purpose?

These questions and others are being researched by top flight specialists in many different fields as you, the reader, proceed to employ his suggest survival methods contained in this latest document attributed to Omar 'The Great Scholar'.

Brief discovered documents trying to answer such questions are discussed in the following paragraphs. Please read same with an open mind and employ common sense when you do.

8.5   Sayings Attributed to Omar

Actual documented sayings attributed to Omar, 'The Greatest of All Scribes', 'Fountain of Wisdom', 'Holder of Knowledge', as uncovered and translated from various ancient languages by experts now follow for the reader to ponder. Doing so with some care is strongly suggested.

The reader must also deal in this section with explanations found in front or behind Omar's sayings as they occur for addition guidance, and aid to there origin, condition, and placement in Omar's body of documented prose and poetry works.

Another discovery has been made in the writings at the "dig site", which, this time, seems to be a work on philosophy that is very profound even to this very day.

Its author is identified as 'Omar the Mournful', who some scholars have named, 'Omar the Unbelievable', because of his wide ranging intellect, seems to be discoursing to his students on the subject of honor. Remember we have discovered that he was a teacher of great merit at the University of Bullistand, located near present day Tyre, in southern Turkey. We also know that teaching was just one of many pursuits.

Omar

8.6 Omar's Concepts of Honor

Following, below, are the best translations available from the ancient scrolls on Omar's concepts of 'Honor' as found in what is believed to have been the library of the University of Bullistand, where Omar apparently taught.

"Honor? You ask what is honor?

It is not what you think! That I can tell you. For small minds distort it into a thing. Clever minds use it for their own ends. Devious minds warp it into a non-recognized entity for their own use. Rulers use others honor to gain more power for themselves.

Those who use dishonest weights and measures wear it as a covering, while they take advantage of the week, helpless, and unknowing among you. Money lenders hide their dishonesty behind it. Most use 'honor' to gain their own ends, in a fashion that makes Allah Himself weep for us all.

Now I will tell you what honor is.

It is the never ending pursuit for what is the absolute right, in all walks of life, by ordinary simple people, every day.

It is a mother loving her child.

It is a father providing for his wife and children.

It is a warrior protecting the weak.

It is love of all that is right.

It is the doing of good for others while we are living.

That is 'honor'."

The ancient document seems to be a bit difficult to translate from this point, but the translators are still working on the problem. In any case, perhaps our current leaders around the world should apply a bit of Omar 'The Insignificance's' 'honor' to their decisions, so that perhaps the world would be a better place.

8.7 Omar's Sayings on Various Topics

Most of the following quotes from Omar were found at a site just located east of the city of Haifa, located in modern Israel's northern area. The exact location is being kept secret to protect what may prove to be a "mother load" of Omar's writings.

"Life for most is like wandering in the great Empty quarter, the vast desert that has nothing in it. Even after the wander has passed by, nothing has changed."

"For the fortunate few life is like being in a constant oasis, full of green growing plants, flowing fresh spring water, shade in abundance. cooling winds, and at night a sky bursting with the brilliance of stars beyond count."

"Power is toxic, similar to the desert asp. Avoid it."

"Work is necessary, sweat results. Wear is its companion. Wisdom is its younger twin. Embrace it. A price exists on everything."

"A wise- slave keeps as much distance between himself and his master as possible. Why? His whip cannot reach into forever."

Note: The above were taken from Omar's "The Slave's Handbook" and was quoted from his rules of survival chapter.

"One should strive to win. One should always be prepared to lose. And, one should remember, the race does not always go to the swift camel, but to the one who seems to finish first."

Note to the reader: The above is from a fragment of Omar's management teachings to his owner.

8.8 Omar's Sayings for Slaves

"When the master is less a person than his slave, and doesn't know it, a wise slave keeps silent. For if a slave's abilities are more than his master's, envy will result, which brings the whip. And, if a slave's abilities are less than the master's, ego will result, which brings the whip. Regardless, the whip will come!"

The slave keeps his silence, and must meld into garden wall, does as little as possible, to catch his master's eye."

"A wise slave will say what his master wants to hear, when his Master wants to hear it, and how his Master likes to hear it. At times this approach will be questioned by the Master, but the wise slave will point the Master toward another. This other one, as innocent as a white lamb, will then take the master's wrath. This is known as the 'Kiss masters ass technique'. It seems to work on most Masters."

"The best slave from the Master's view sees nothing, hears nothing, tells nothing, but does (so it seems) everything when he is told, how he is told, as he is told. This is known as the 'I am your boy' technique."

"Remember the great rule: 'A slave only exists to serve.'"

"The loaf of bread belongs to the Master. Crumbs to his slave."

"Life is lived by the Master. Slaves just exist."

"A slave cannot outlast his Master. The rules of the game prohibit this. Remember this."

Note: The above was translated with great effort by a team of mid-East linguists working with only fragments of documents over a thousand years old. However, they are sure that the essence of their work on "Omar's Handbook for Slaves" was correctly translated.

"Slaves are just fodder. Grist for the mill." - from Omar's manual for Masters.

"Keep silent. Keep low. Keep down. Let the other slave get his head conked, not yours."

"Always pass the blame. Never accept same. Always have a 'fall' slave. Its very convenient for masters."

"Never, I say again, never allow any slave to take credit for any success of any type or kind."

"Remember, my son, slaves come and slaves go, but masters last. It is in the rules of slavery." - from Omar's Last Memoirs (translators think).

"Get rid of a slave who has a mind of his own. After all, slaves have no mind as it also belongs to his master."

"Slaves think? Property can never think, my son."

"Who gets credit for ideas? Why of course, the master does. Always the master."

8.9   Hints on Omar's Character

Note: Scholars have found another set of documents on ancient wisdom apparently written by "Omar the Least".

This set is now undergoing much discussion as to content, as well as the context of the times "the Insignificant" lived in which he wrote his various "books" and his yet to be published work "Weightless Thoughts on Heavy Meanings", otherwise know by scholars as "The Book of Nothing on Everything".

Interpreters seem to be having a trying time, as "the Most Humble" often wrote in a form of poetry, or at least poetic scribbling on the most profound subjects concerning mankind. The following are more examples, apparently taken from one of his many lectures at the unknown "School of Wisdom" he lectured at in Unstandbool, in Kirstanand (so scholars now believe).

"Never give thought of disdain or even disquiet of the smallest nature to any doing good for you, to you, with you, around you, or yours; only show the highest admiration and humble appreciation for all goodness bestowed."

Some scholars translate the above passage, as follows:
"Careful that you don't shoot yourself in your own damn foot!"

Note: However, not likely, as firearms came a millennium or two later.

Note: A description by Omar's supervisor showing one of his actions as translated from the same hidden dig, apparently a vision into the future, now follows.

"Omar just ran into my office in a state of great excitement. He claimed that he was awakened about dawn by a 'hopping hopping' noise outside his window, got up, looked, and saw (much to his astonishment) a giant white rabbit carrying all sorts of things. It very quickly just disappeared as the sun's rays began to hit it (not visible in white light, apparently).

Omar stated:" Never in my long life have I, Omar, the most least of all, servant of Allah, the Most Merciful God, scribe to David the King, Ambassador to the Pharaohs, Citizen of Bullustand, and last among your servants, have ever witnessed such a sight. This is worthy of contemplation, great thought, and meditation of the deepest kind, and perhaps a new chapter in 'The Chronicles of Life' (a great book of philosophy, by Omar the Lessor)." With that, he wished that I forward a "Happy Easter to the Young Queen of Youthful Womanhood", and left my office."

Note: Scholars have yet to identify the "Young Queen", nor have they yet identified just who Omar's supervisor was during his teaching at the university. Mysteries abound about Omar.

8.10 Omar's Birthday Disclosure

Here is what Omar has to say about one of his life times he claims to have lived as disclosed from another "digs" findings.

"My memories of that day so long ago are just blurred with the fog of time. Why I was sent this existence on the time line of infinity has always been a mystery to me. It brings into question the final meaning of life, time, soul, mind and my relationship to Allah, blessed be His Name, Himself.

However, I do know that life began for me on that "Day-of-Days" many centuries ago (again he did not say his age). My mother was a genteel soul. Always helpful, always understanding, and near whenever I needed aid and comfort. My father was a "man's man", and served the Grand Mufti of Puttinstand with great honor and bravery.

Raised was I in a caring household, filled with servants, with no opportunity denied for gains, knowledge, and experiencing all the good one's life has to offer. Perhaps, on this modern scale of yours, I would be considered a child provide all, including great material wealth, which seems to mean so much to your people.

Alias, on each one of my multitude of "Day-at-Days" I reflect back across spans of years, more now than I can count, or even clearly remember. I and wonder at the enormousness of my life: student, scholar, warrior, worker, husband, father, philosopher, writer, singer of psalms, servant to the mighty rulers, and so forth, and cannot but seemed perplexed.

I ask myself, "What does it all mean?". And I have no answer. Silence rules. Even on my "Day-of-Days". Perhaps you know what your "Day-of-Days" means to you, and to Allah. I can only say, (and here I directly quote Omar), "What is written is written. It cannot be changed. It is Allah's will."

And Omar left us, with this, his "Happy Birthday, my son." salutation.

Omar wants me, the editor of this guide, to communicate his message of congratulations on July 4th as follows:

"Omar, this least worthy humble personage, begs to be allowed to wish all American citizens a greeting of joyful celebration on this holiday known in your country as July 4th, Independence Day, from the British oppressors under the leadership of the despot King George III.

May victory always be yours, and Allah (bless His Mighty Name) always shower His grace upon this magnificent freedom loving country of yours. I, Omar, the Meek and Lowest of All, beg you to accept this poor tribute from my trembling lips."

I tried to tell him a simple "God Bless America" would do, but he insisted.

8.11 Omar's Tribute to His Wife

Omar wrote the following, as translated from a "dig", and it now follows:

"Many centuries ago I was a great scholar, poet, writer, and teacher. Life held many wonderful lessons, and treasured ideas. Contentment was my lot, and nothing was needed nor wanted of material things. Yet I was not happy, not really, until I met my wife.

She was at woman of a thousand wonders. A being of incalculable worth. One who always brought happiness to everyone. Her smile alone would turn a dark night into a bright day. Her mere presence was magic like a good genie's, granting a wish of happiness at every turn.

She gave me children/ and the grandchildren, the nectar and essence of our existence. She gave me love, the greatest of emotions, the core of life. How happy I was, and still am, in her glorious presence. How lonely, life is empty without her with me.

God, being lonely, alone, is a terrible punishment. Forgive me, I beg."

With that, Omar (the Most Insignificant of All) got up, gathered his meager belongings, and left without a glance.

8.12 Omar's Statement on Man's Greatest Achievement

NOTE: Associated Press just announced a new translation from an ancient document attributed to Omar 'the Lonely and Lost' as its author. Remember that this man was a great scholar, teacher, scribe, and surely a author of some of the psalms attributed to King David and some of the proverbs attributed to King Solomon. Indeed, one of the ancient world's greatest thinker's, but very unknown until now.

The translators of the ancient dialectics think Omar is lamenting being abandoned by his beloved wife, best friend, life companion, mate, and mother of his children. At any rate, here is another part of Omar's thoughts that translating scholars call "Omar's Lament".

"My students, you ask what is a man's greatest achievement?

The answer is simple, clear, and easily seen by all.

It is his love of his woman, that part that completes his existence, makes him a whole being in the eye's of Allah Himself. Without such a completion, rest never can be found in this life.

Never can a man be content, nor fulfilled, as happiness is just a hollow shell, and thought itself can never be satisfying, nor food taste, wine be enjoyed, and the universe seen as it should be without a man having the love of his woman.

Remember, that being alone is the most awful of conditions that a man can encounter.

Indeed, it is worse that starvation, imprisonment, injustice, and any of the other deadly sins we are to encounter in this meager existence. And a man without his woman is in such a sorry state. Pity him.

For nothing can soothe his hurt, heal his pain, nor relieve his soul. Nothing!!"

9.0 School Day of the Future

What follows is a potential school day in the near or middle future of this 21st century. Omar apparently wrote this as he gazed into the future. The reader should use great imagination as he envisions a school day of the future.

"Once upon a time, in the not too distant future, will live a very small little girl, named Ami. For sure she was small, because she could sit in tea cup. And for sure she will live in the future, because her best friend was a mechanical watchdog name Wolf-Wolf. In fact, her mother got her Wolf-Wolf; so as to always have her safe. No harm could come to Ami, as long as she was watched over by her Wolf-Wolf.

One day, a winter day, because it was cold and the days were dark, Ami and Wolf -Wolf were walking in the school playground (in the near future kids can bring their pets to school), looking for someone to play with. Jack came flying by, on his sky slide.

Now remember, in the near future toys will be 'smart', and able to 'think', and all such toys will never allow anyone to get hurt during playtime.

Jack yelled down at Ami, "Hi, Ami". Come on; let's play sky soccer! Get your sky slide on, leave Wolf-Wolf here and join the fun." Ami turned to her 'dog', stay here. While I go play sky soccer.

She hit the button on her personalized caller, and ordered her sky slide from her game storage locker in the gym. It responded at once, following the radio signal of her personalized caller (similar to today's 'beeper'). She jumped on it, and joined the game.

Now sky soccer is played almost like the game soccer, except it has the added dimension of height, not just a field to run up and down upon. Goals are still much the same, except instead of a net that stays put, the net moves up and down, at the end of playing each "cube" volume, so as to make the game more varied, as play progresses.

Ami loved to play sky soccer. Why? Because she was small, but very fast on her sky slide, and just about the same size as the 'floating soccer' ball. It contained an antigravity generator, a directional finder, and a self-referee 'thinks' module, So all the rules of the game were in the floating soccer ball itself: and it called its own game, as both sides played. Fair too.

Ami could ride right behind the floating soccer ball, kicking it from her sky slide, and always make goals, n0 mater how the goal moved itself. Because of all of this, Ami, was the best player at school of this game. Her side almost always won. Everyone cheered whenever she made a goal, as her small size and quick speed were wonderful to see.

So the game started, with everyone having a good time. Just as Ami was about to score her third goal, the bell rang. Her teacher, Ms. Know-Everything, called out over the playground speakers. "Back to class, please". So everyone stopped, landed, and ordered their toys back to gym storage. Wolf-Wolf was glad. He never did like flying about himself, although he could, but kept that a secret from Ami.

All of the children, including Ami, headed into the school, back to class. However, in the near future, the classroom is far different than in 1998 (the year of this soon to be true story). In Ami's classroom, space was controlled by Ms. Know-Everything. With the push of a button, the roof opened up, for her class on clouds, stars, planets. Or any topic needing lots of space. At least it seemed to, but actually a holographic projection system controlled by a very fast computer system, with links to databases containing all knowledge (up to yesterday, of course) provided the illusions MS. Know-Everything to teach from.

As Ami listen to the lesson on planet Mars geography (people live on Mars in the near future, several thousands of them, and on our moon too), she daydreamed of going to Mars's north pole, and ice skating on its polar ice. She could do a 6-trlple turn, or a giant Mar's split, and other ice skating maneuvers that cannot be done on our planet, both due to gravitational pull here being so much greater.

Ami also thought of going to explore Mar's two moons, just for the fun of it. On Mars she would be able to lift more weight, walk further, and have more days in a year than on Earth. What fun! Just then, Ms. Know-Everything stopped her sky series lectures for the day, switched the holographic system to natures insects, for a giant grasshopper appeared in front of the classroom, next to Ami's learning cube.

As the grasshopper was being rotated about for a better view (Sammy, in the back of the room, was always asking for a better view), Ami had to forget Mars, and began to concentrate on the grasshopper as nature's "bug" for today's lesson in insects. The strength, exoskeleton, number of legs, types of eyes, and more facts than she wanted poured out of the database, visually appearing in an automatic fashion, with floating labels pointing to the insect's body parts. Ugh, Ami said to herself, who cares!

Finally, the natural science illusion was stopped, and suddenly beautiful flowers appeared, in a garden setting. It was students expression time. Ami reached for her electronic paint pointers. This she would enjoy (Sammy did not, as a groan was heard from his learning cube). With deft strokes Ami filled in her partition of the garden setting. Each of her classmates (only those interested) were doing the same. When finished, back into the database of class accomplishments it would go, so parents could see what their class had accomplished in the arts instruction. As she painted, Ami listened to Ms. Know-Everything talk about color, shades, tone, texture, shapes and other art terms. Ami knew that she enjoyed this part of her day.

Now, 'free time' was at hand. All students that bad been paying attention, done their work, and caused little problem in the class were allowed 'free time'. Ami decided to write a poem for her mother, who was at work at the local medical complex. So Ami began to think, and to write. Here is her poem, after much thought, sent over the Speed Net to her mother at work.

My Mother- My Treasure
by Ami

"Smiles and hugs-come from you,
Kisses with love surround you,
Life worth living, forgiving are you,
And who is you, but my Mother.

Better than treasure, more than can be measured,
Thoughtful, kind, and helpful too,
Worth more than diamonds, gold, and anything else,
Best for me, My Mother."

With the end of her poem for the day, Ami and the other students had to work math problems. The usual groan was heard from Sammy, because he always groaned at any work assigned by Ms. Know-Everything, As the classroom was prepared for math, visual aids appeared. Blackboards, light pointers, three dimensional equations for reference, and

yesterday's math lesson (recalled from the database, of course) appeared in each learning cube.

Today's lesson was in basic volumes, calculating how much a space box would hold if it was so many meters in height, width, and depth. Wolf-Wolf woke up, as he had a advanced math chip in his computerized brain, but he went back to 'sleep', since Ami did not need his help. After several problems, the math session ended.

Now was story time, Ami was very intent, as she wanted to know more about ancient schools, and how subjects were taught long ago. One of her ancestors was a teacher, so she remembered her father saying once. Ami wondered just what school was like long ago, as Ms. Know-Everything began reading from her teacher's lore book, another school day came slowly to a close.

10.0 A Letter to Heaven

Completely unknown to me until after her death is the following letter, written by my wife to her own mother, years after her mother's passing, found as I was going through my wife's papers. It was so soul touching in so many uncountable ways that I felt that it should be included in "Omar's Guide to Survival"

It illustrates what attributes it takes to make one's life not only as a survivor, but as a God given wonderful growth experience for body, mind, emotions, and soul.

Contained in this most heartfelt letter of tribute are found the following:

1.) thanks freely given, and openly expressed, for all help received
2.) understanding of and at all seasons of living, is necessary
3.) love is the most essential human emotion to give and to receive
4.) love well spent is never forgotten
5.) family is essential
6.) laughter is a survival necessity; always
7.) prayer to God is always needed for survival; always
8.) respect for mothers; respect for others
9.) memory of all kindness received, small or large, is the salt of life
10.) gratitude of the deepest sort should be expressed

Other additional attributes are contains in the letter for the reader to discover, such as pain, is part of survival lessons for the reader.

Simple ideas wonderfully expressed are often heard, read, or found in the most unexpected places, at the most unexpected chaotic times, as each person's time line in life progresses.

Dear Momma,

I just had to write this letter. Maybe some kind person will read it to you. Last night I lay in bed trying to count the many thanks I owe you. Somewhere along the way, I fell asleep. Well, here it goes.

Thank you sweet lady for being there during my helpless years. Now that I've been a mother and teacher for several years, it is so clear to me that many of our pains we carry as adults are throwbacks to our childhood.

Thank you Momma for being there for me when I was an obnoxious, vain teenager. Never did I understand how hard those years can be for Moms. We know and understand nothing.

Finally, thanks Momma for being the close and loving mom I needed so badly when I

married and started my family.
It seems like yesterday
when I would come visit
and we would sit up late
at night laughing at
things that happened in
our family.

Now, I'm there for you
in spirit. At night I talk
to you through my prayers
and feel you beside me.
Every night I say, "Good
night my well loved and
respected Momma."

You're always close - at
least that's how I feel.

This is just a way to
say that all your love is
well spent. I treasure every
memory in my heart.

Your loving daughter,

Laura

Happy Mother's Day!

11.0 A Wife's Love

This following letter illustrates a wife's uplifting effect upon her husband by sending a short meaningful note sent to him while he was on a business trip. This type of letter
may be short in length, but it's message lasts in ones heart into eternity.

Talk to you tonite.
Remember - you're never
totally alone, you're in
my thoughts & heart day
by day-especially at nite. I

love you.
    Always yours,
        Laura

12.0 Mom's Weight Loss Letters

The following letters contain very good advice from my mother on losing and maintaining your weight for good health. She had a real problem later in life, and had to lose a great deal of weight, as her heart began to fail. She died at the age of 82 in the year 2000 from heart failure in Austin, Texas. Perhaps being the major care giver for my father for several prior years hastened her demise.

love you.
always yours,
Laura

Dearest Son:

To loose weight is one of the hardest jobs you will ever do. Don't give up. Your family is all behind you and wish you the best. The best advice I can give you is ① if you are taking water pills be sure to increase your potassium in take or you will have leg cramps, boy do they hurt. Eat an extra bananna and plenty of orange juice each day. Lots of fruit for vitamin C and a good vitamin supplement. Remember meat gives your blood strength, beef or pork roast without a lot of fat. Pinto beans and rice are especially good. Allow yourself a good desert jello (sugar free) and Blue Bunny are good. (I like the fat free Fudge Bars.)

Weigh yourself each morning when you get up. Eat one fish meal a week. (Luby's)

Most of all be happy. You and Laura deserve to be happy, you have accomplished a great deal. I am so proud of you both.

Love Mom.

P.S.
Call me let me know how you are doing. Judy talked to Cherry Huffin to day, small world. They are re-modeling their house. Mr. Huffin is 91 years old.

3-14-2000
Austin, Texas

Dearest Son:

So good to talk to you on the phone. Glad everything seems to be going fine at your home. I love it here with such fine neighbors. They look after me. I am so proud of you + Laura and your family-great job you both have done.

Loosing weight is awfully hard to do and to keep it off is harder. The calcium tablets I told you about are called Calcium + Magnesium all natural. I get mine at H.E.B. Try them if you don't like them not much lost. They helped my leg cramps.

We all seem to be doing fine here. Hope to keep it up. Call when you can.

Love You All.
Mom

13.0 A Granddaughter's Assignment

At 16 years of age, and in the 11th grade, my granddaughter wrote the below note to her English teacher. She has been diagnosed with a serious sleep disorder (narcolepsy), caused by certain brain neurons "misfiring" at random times, resulting in loss of control of her body. It is a life long condition, unless such a young person "out grows" it.

Her choice of words in describing its effects, and her reactions to same, contains very mature thought provoking responses at such a young age. Most adults would not be able to match her spirit, strength, intellect, and determination in a similar situation.

She makes me very grateful and extremely proud to have such a granddaughter. Her generation is indeed the healthiest in body, mind, and spirit I have ever witnessed, and the most gifted, in spite of the world we older generations have created for them, to live in, with all its vast sets of unsolved problems.

God bless her, and her entire generation, as they solve the problems we have left.

Addison Jackson
Jana Reid
AP English N - 4
17 September 2015

Lifetime Goals

"You have the same disposition your brother had." I move my eyes to my teacher's, with the same void daze that makes up my presumed "disposition". "He always seemed disinterested, bored rather. But Austin emanated brilliant thoughts. Always."

My cheeks flush, and I am somewhere in between bashful and abashed at such a particular observation. I rack my brain for whatever brilliant thoughts I was previously thinking, and can only find the urge to sleep next to the fog that fills the rest of it. My dosage must be low, I think to myself: Sometime after the fact of my diagnosis with narcolepsy. I guess I subconsciously started using the word "dosage" interchangeably with focus, or the allotted time I had until my body would power down again. I am chronically tired. How much I accomplish in my everyday life is a function of my ever changing state of wakefulness.

My sleep specialist described the condition to me as an analogy in which the brain is a stick-shift car, and the sleep cycle moves in gears. A normal person would have the appropriate sleep latency and intervals between sleep stages up until REM, or the deepest state of sleep in which we dream, at about 60-90 minutes after hitting the gas pedal 1,2,3,4, cruise. A narcoleptic's cycle shreds through this pattern and peels off into the distance within minutes or even seconds, not to be found mild, it fluctuates randomly. (5,2,3, 1, wake up, 5, etc.) This backwards event of the brain's control center being out of tune with the rest of the body is represented in symptoms like sleep paralysis, the phenomenon of conscious dreaming. Subjects fall into REM way before the mind has time to follow the proper procedures for good sleep, and the body literally falls into the paralysis that prevents you from moving during sleep. Prior to my understanding of what it was, this happened to me repeatedly. The delusions are surreal and there are no limits to your senses in this state. It's difficult to capture what it's like to be surrounded by black alien shadows or electrocuted by the Messiah (one episode in which I went into what felt like ventricular fibrillation), let alone be paralyzed from your toes to your eyelids. I don't know what the average person classifies as normal bed activity, but I am boggled and mortified by what a sleep disorder could be responsible for. For a long time in the past and sometimes now, bouts of paranoia and depression concerning my wacko skizzo sleep problems creep up. However, I've been enlightened and assured of at least one thing, I can devote my life to.

I want to find a better route of correction fur the ones who could end up in an unorganized life like mine, and one that isn't composed of amphetamines by day and rookies by night. This is sadly the best treatment available to narcoleptics today. It's unsettling to realize most of my energy is artificial, and the more I up the ante in terms of medication, the harder it will be to do without it. I once held a high place in regards to my teacher's standards because I genuinely enjoyed work, but at this point in high school, one error seems to exponentialize.

When it's hard to keep up with the common tasks of life, any want for excellence is made dreamlike.

There are things I aspire to do. I want to march in drum corps and I want to raise children, but it's not those dreams alone that motivate me. I have the dream to do these things by forcibly placing mind over the matter of chronic illness. I will find the threshold that is supposed to separate me from greatness, and break it. Every action is the beginning of a chain reaction that wouldn't stir without it. My life is to be a small action towards a big movement for better living long after I'm gone.

14.0 FINA XXXXXXXXXXXXXXXXXXXXXXXXXXXXXXXXX XXXXXXXXXXXXXXXXXXXXXXXXXXXXXXXXXXXXX

Printed in the United States
By Bookmasters